Surface Chemistry of Carbon Capture

Surface Chemistry
of Carbon Capture
Climate Change Aspects

K. S. Birdi

CRC Press
Taylor & Francis Group
Boca Raton   London   New York

CRC Press is an imprint of the
Taylor & Francis Group, an **informa** business

CRC Press
Taylor & Francis Group
6000 Broken Sound Parkway NW, Suite 300
Boca Raton, FL 33487-2742

First issued in paperback 2021

ISBN-13: 978-1-03-208555-5 (pbk)
ISBN-13: 978-1-4822-9968-7 (hbk)

Publisher's Note
The publisher has gone to great lengths to ensure the quality of this reprint but points out that some imperfections in the original copies may be apparent.

Visit the Taylor & Francis Web site at
www.taylorandfrancis.com

and the CRC Press Web site at
www.crcpress.com

Contents

Author

K. S. Birdi joined Standard Oil of California, Richmond (1957–1958), after graduating from the University of California, Berkeley, 1957. He joined (chief-chemist) UniLever Copenhagen, Copenhagen, Denmark (1959–1966). Later, he was appointed associate professor (1966–1985) and research professor in 1985 (Technical University, Copenhagen; Nordic Science Foundation), and was then appointed, in 1990 (retired in 1999), to the School of Pharmacy, University of Copenhagen, as professor in physical chemistry. Professor Birdi has published various books related to surface and colloid chemistry: (*Handbook of Surface and Colloid Chemistry*, 1st Ed. (1997), 2nd Ed. (2003), 3rd Ed. (2009), and 4th Ed. (2016), CRC Press, Boca Raton, FL, USA; *Surface Chemistry and Geochemistry of Hydraulic Fracturing*, CRC Press, Boca Raton, FL, 2017). Professor Birdi has participated in various national and international research projects related to North Sea oil explorations and European Union joint projects.

Introduction to Surface Chemistry of Carbon Capture

1.1 INTRODUCTION TO SURFACE CHEMISTRY OF CARBON CAPTURE

The Earth is surrounded by atmosphere (air), and water is found in great oceans (covering almost 75% of the surface of the Earth):

ATMOSPHERE

. . . .

. . . .

. . . .

EARTH—OCEANS

Furthermore, the Earth—atmosphere—sun system

SUN—ATMOSPHERE—EARTH

composes a configuration that mankind has admired and studied for many centuries. Ancient religions and civilizations have worshipped certain specific natural elements for thousands of years, and some religions still worship them. These elements were

Air (atmosphere, wind)

Water (rivers, oceans)

Sun (light, heat, radiation)

Fire (flame, combustion)

COMPOSITION OF AIR (ATMOSPHERE)

The composition of air (or the atmosphere), as known today, is a mixture of different gases (near the surface of Earth):

nitrogen (N_2: 28 gm/mol) (78%);

oxygen (O_2: 32 gm/mol) (21%);

argon (Ar: 40 gm/mol) (0.9%);

carbon dioxide (CO_2: 44 gm/mol) (0.04%);

water (H_2O: 18 gm/mol) (vapor) (traces);

other gases (hydrogen, etc.) (traces).

The global average temperature of the Earth is primarily regulated by the heat received from the Sun (Appendix C). Furthermore, the energy (heat, radiation) reaching the Earth from the Sun has to pass through the atmosphere surrounding the Earth. This means that the sun's energy (radiation) has to interact with

gas molecules (and dust particles) in the air (atmosphere)

reflection by air/clouds

reflection from Earth, trees, and ocean surface

The heat from the Sun (the surface temperature of the Sun is 5500°C) is transmitted by the infrared wave length of the radiation. The infrared is absorbed by some gases found in the atmosphere (e.g., carbon dioxide, methane, etc.). These gases are called *greenhouse gases* (GHGs). The absorbed infrared energy is reflected in all directions, which increases the temperature of the planet Earth. In the present text, there is only interest in **carbon dioxide** and its interactions with the temperature balance of the Earth.

The concentration of carbon dioxide (CO_2) in air is reported to be increasing over the past century (ca. 2 ppm/year) (IPCC, 1995, 2005a, b, 2007, 2019; IEA (International Energy Agency) 2011, 2012, 2013, 2014a, b, c,

2015a, b, c, d, 2016e; EPA Handbook, 2011; Dubey et al., 2002; Sanz-Pérez, 2016; Rackley, 2010; Dennis, 2014; CSIRO, 2013; Albo et al., 2010; Oh, 2010; Hinkov et al., 2016). This has been attributed to fossil fuel burning (combustion) (anthropogenic CO_2) (Leung et al., 2014; McDonald et al., 2015). Furthermore, the increase in population and increased energy demand leads to increase in anthropogenic CO_2.

This book relates to the carbon (i.e., CO_2) capture and **surface chemistry** properties of **carbon dioxide** (CO_2), especially its connections to the change in temperature (average temperature) of Earth. However, it is useful to describe some general remarks about the atmosphere (air) and its characteristics.

The composition (and temperature and pressure) of air varies with height over Earth. In other words, when one considers the composition of air, the height has to be considered. The gradient of composition of air (i.e., variation of composition with height) might be dependent on time scale (and space). Similarly, water (e.g., oceans, rivers, lakes, drinking water) contains different salts (sodium, calcium, chlorides, sulfates, etc.) that are known to interact with some components of air (for example, carbon dioxide, oxygen). Mankind has been aware of the essential role played by these elements for the existence of life on Earth. Even though millions of miles away, the Sun provides heat to the Earth. Besides heat, the Sun also emits other kinds of energy to Earth, for example, radiation, ultraviolet light, and so on. The atmosphere surrounding the Earth has four distinct layers (Saha, 2008; Lang, 2006):

0 to 10 miles (0 to 16 km: temperature range, ca. 20°C to −50°C) **TROPOSPHERE**: This is the region of human activities.

10 to 30 miles (16 to 50 km: temperature range, ca. −50°C to 0°C) **STRATOSPHERE**: The ozone gas (O_3) layer, which absorbs the ultraviolet (UV) radiation from the sun, is found in this region. The temperature is higher at higher altitude, owing to the absorbance of ultraviolet radiation from the Sun.

31 to 53 miles (50 to 80 km: temperature range from ca. 0°C to 90°C) **MESOSPHERE**: In the mesosphere, the temperature decreases as altitude increases.

53 to 75 miles (80 to 200 km: temperature from ca. >90°C) **THERMO-SPHERE**: In this region temperature increases with altitude, owing to absorption of highly energetic solar radiation.

This clearly shows the complex sun—atmosphere system and the interactions relating to temperature on the Earth.

Further, mankind has used **fire** (with respect to food, combustion processes, energy (electricity, mechanical), transportation) in many different applications and technologies (both directly and indirectly) for many hundreds of years. It is also recognized that all these elements are essential for the existence of life on Earth. Fire or *combustion* of fossil fuels in the modern age is described as

COMBUSTION: FOSSIL FUELS + OXYGEN (FROM AIR) = FIRE (heat, electricity, mechanical energy, etc.) (CO_2 produced)

This process (i.e., the production of carbon oxide) is anthropogenic and is currently related to the world population (e.g., food and energy demands). It is thus seen that the above process of combustion adds CO_2 to the air. At the same time, it must be stressed that carbon dioxide in air is essential for the existence of life on Earth.

PHOTOSYNTHESIS AND CARBON DIOXIDE CYCLE

The most significant process with regard to life and all kinds of living species on the Earth is the supply/production of food. The latter is singularly provided by the complex system that is dependent on the photosynthesis process (Paoletti et al., 2002; Raschi et al., 2000):

CARBON DIOXIDE (CO_2) + SUN SHINE + WATER < PHOTOSYNTHESIS > ALL KINDS OF PLANTS/FOODS/CARBOHYDRATES

Each gas component in air is known to have a specific role in the life cycle on Earth. For example, even though the concentration of carbon dioxide (CO_2) in air is at present only 400 ppm (0.04%), it provides all **carbonaceous food** to life. Food is obviously the most necessary product for the existence of life on Earth. Actually, the photosynthesis process is the opposite of fossil fuel combustion, and CO_2 is captured in plants and so on. It is important to mention that all the carbon atoms in plants on Earth are supplied by the **CO_2** in air (via photosynthesis). Furthermore, the food is digested by all living species on Earth by the metabolism process (Appendix C):

The CO_2—food—metabolism cycle:

CO_2 in air (photosynthesis) >>>> Food >>>> Metabolism (exhale CO_2)

This shows that in the life cycle on Earth all the carbonaceous components are solely provided by CO_2 in the air (EPA Handbook, 2011). Actually, it is clear now that all these natural elements are interrelated and are basic necessities for life on Earth. The characteristics of the different gases in air (atmosphere) for example:

Oxygen (O_2: 21% in air) is essential for all living species (as a source of metabolic and other reactions, for example, oxidation). Furthermore, oxygen is an essential component of all combustion reactions. In all combustion processes, when fossil fuels, wood, etc., are burned in air (containing 21% oxygen, O_2) the reaction is as follows:

Fuel (consisting of carbon and hydrogen molecules): $(C_x H_y) + O_2$ (gas) $= CO_2$ (gas) $+ H_2O$

CO_2 (even though present at a very low concentration, ca. 400 ppm) supplies all the food (photosynthesis process) for all living species (as well as all the organic molecules, which mainly consist of carbon atoms). Most significantly, it is important to mention that the concentration of CO_2 in air must be sufficient to support the need for plant growth on the Earth. Therefore, one will expect that there must be a **minimum** concentration of CO_2 (in air) in order to sustain life on Earth (Appendix C).

Nitrogen (N_2: 78%) Even though nitrogen does not interfere directly in the life cycle, it is converted to ammonia (NH_3) after interaction with hydrogen and is used as fertilizer for growing plants (food). This means that the increasing demands for abundant food production globally (related to an increase in world population) need fertilizer to assist in this demand.

The different gases in air exhibit different physicochemical properties (as regards to solubility in water, absorption of infrared light, etc.). As regards the solubility in water, it is found that the solubility of oxygen and nitrogen is very low. However, CO_2 is soluble in water (i.e., in oceans/lakes/rivers) (IPCC, 2007; Rackley, 2010). Hence, one finds that there is a very large amount of CO_2 in oceans, which is in equilibrium with air.

CARBON DIOXIDE IN AIR === (EQUILIBRIUM) ===
CARBON DIOXIDE IN OCEANS

This means that any changes in the concentration of CO_2 in either phase (e.g., air or water) will lead to a corresponding change in the other phase (Chapter 1.3; Appendix C).

CARBON CAPTURE AND STORAGE (CCS)

The **purpose of this book** is to address the present technologies available to control the CO_2 increase (due to anthropogenic CO_2 emissions). The CO_2 (as gas) can be captured/sequestered by different processes (Chapter 5) called carbon capture and storage or CCS processes (Chapter 4) (figure 1.1).

CCS (carbon capture and storage):

Carbon dioxide (CO_2) <> Capture process <> Storage of captured CO_2

One finds in literature that carbon (i.e., CO_2) can be captured (from flue gases) by various methods (Suzuki, 1990; Vargas et al., 2012; Krishna & van Baten, 2012; Yu et al., 2012; Bolis et al., 1999; Hinkov et al., 2016; Rackley, 2010):

gas/adsorption on solid

gas/absorption in fluid

other different technologies to capture carbon (CO_2)

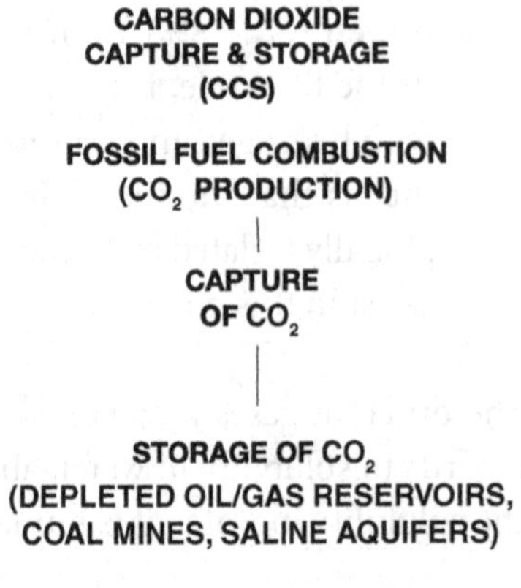

**CARBON DIOXIDE
CAPTURE & STORAGE
(CCS)**

**FOSSIL FUEL COMBUSTION
(CO_2 PRODUCTION)**

|

**CAPTURE
OF CO_2**

|

**STORAGE OF CO_2
(DEPLETED OIL/GAS RESERVOIRS,
COAL MINES, SALINE AQUIFERS)**

FIGURE 1.1 Carbon capture and storage (CCS) schematics.

These studies are described in Chapter 2.

It is also important to mention (briefly) other related phenomena before describing carbon capture and its surface chemistry aspects. Mankind is increasingly interested in understanding and controlling the various phenomena surrounding the Earth (such as pure drinking water, clean air, average temperature of the Earth). The average global temperature (this term is very complex, as one might expect) on Earth is related to different energy sources/sinks and equilibrium between the latter.

SOURCES OF HEAT AND SINKS IN EARTH

In general, the Earth's heat balance with its surroundings (energy sources or energy sinks) that need to be considered are as follows (Appendix C):

SOURCES OF HEAT

Sun (major)

Mankind combustion technologies (oil + natural gas + coal)

Diverse sources of energy (plants, forests, storms, etc.)

Earth core (temperature is very high)

SOURCES OF HEAT SINK

Evaporation of water (from oceans, lakes, rivers, etc.) to make clouds/ trees (which reflect the Sun's heat)

DYNAMICS OF TEMPERATURE OF EARTH

It must be mentioned that there is no parameter (e.g., temperature, pressure, wind velocity, etc.) that is static at any point of observation on the surface of Earth. All parameters are varying with time and space on the Earth. For example, on the average, on a typical day, the temperatures (near the surface) at any place on Earth are higher during the daytime than in the night:

Temperatures at different parts of the Earth

CITY DAY/NIGHT TEMPERATURE (°C)

DATES	*MAY 2018*	*JANUARY 2019*
COPENHAGEN	*−1/−1*	*3/−2*

LONDON	*11/4*	*6/−1*
FRANKFURT	*10/−3*	*3/−2*
LIVIGNO	*1/−6*	*−5/−13*
MOSCOW	*−7/−20*	*−7/−7*
CALGARY	*−6/−17*	*−2/−13*
DELHI	*28/14*	*19/6*
NICE	*12/7*	*14/2*
BUENOS AIRES	*28/16*	*32/26*
MALAGA	*18/10*	*17/11*
NEW YORK	*7/0*	*2/−2*
MANILA	*28/21*	

In addition, the temperature of oceans also changes with day/night/summer/winter parameters. The day/night temperature data merely show the heat (dynamics) the Sun adds to the Earth at different places (and times) on the Earth. These observations indicate that the temperature fluctuations existing at every moment around the clock and the huge magnitude of the Sun's energy (which fluctuates between day and night) reaching the Earth. It is obvious from these data that any analyses of average temperature of the Earth would be rather complex. Especially considering the fact that mankind has no control over what process are prevalent on the Sun and the changes that may have a profound effect on the climate on the Earth.

In the same context, the heat reaching the Earth from the Sun is also different with time and space. The rotation of the Earth also contributes to the dynamics of temperature at any given time or place. Further, the dynamics of temperature is known to create weather turbulence, that is, weather changes (as regards temperature, winds, waves, storms, etc.) due to differences in heat at night/day but also other related phenomena over weeks/months/years.

ATMOSPHERE AND THE SUN

The atmosphere surrounding the Earth is a very important element as regards life on Earth. The composition of atmosphere (air) and its temperature (and pressure) are found to vary with distance from the surface of Earth. The Sun's energy gives rise to heat on Earth (see day/night temperature). The atmosphere (and clouds) controls the amount of heat reaching the Earth in some complicated paths (e.g., absorption, reflection, etc.).

The Sun is ca. 92 million miles (160 million kms) away from the Earth. The Sun's energy (as heat) is found to be enormous as compared to anthropogenic processes (Appendix C). The degree of sunshine reflected by the atmosphere is also a very important element. The degree of reflection by the atmosphere is also dependent on height (from the surface of Earth).

This book relates to surface chemistry (a special theme in general chemistry) (Chapter 3; Appendix A) and its relation to control/storage of CO_2 in the atmosphere (current concentration = 0.04% = 400 ppm). Over the past few decades (especially after the industrial revolution) the concentration of CO_2 has been found to be increasing (Figure 1.2) (IPCC, 2019):

TYPICAL CO_2 (PPM) DATA (YEAR/PPM):

1960/310 ppm

1975/330 ppm

1980/340 ppm

1995/360 ppm

2010/390 ppm

2017/410 ppm

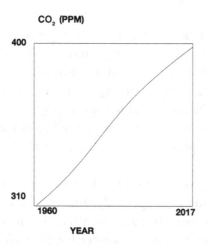

FIGURE 1.2   Change of CO_2 concentration in air (near the surface of Earth).

This observation is attributed to the increasing use of fossil fuels (proportional to increasing population). Currently, global consumption of different fossil fuels is ca. 33% each (equivalent to 100 million barrels/day oil) oil, coal, and natural gas.

At the energy/power plants where fossil fuels are used, the flue gas contains ca. 10% CO_2 as compared to 0.04% (400 ppm) in air. Carbon capture (by application of CCS) can be used in order to mitigate the increasing CO_2 concentration. In fact, it has been recognized that CCS is the only viable technology to mitigate increases in CO_2 and control climate change (Chapter 4). This book describes the technology and surface chemistry of carbon capture and storage (CCS). The main approach is to present a review of CCS technology as found in the current literature.

GREENHOUSE GASES (GHGs) IN THE AIR

Climate change (temperature increase over the past decades) on Earth is related to greenhouse gas (GHG) emission (Dennis, 2014; Yu et al., 2012; OECD/IEA, 2016; Rackley, 2010).

The main GHGs are

Methane (CH_4)

Carbon dioxide *(CO_2)*

Nitrogen oxides (NO_x).

Fluorocarbons ($F_x (H)_y$)

Methane and CO_2 are found in nature as free gases. CO_2 is also produced by all fossil fuel combustion processes. However, NO_x and F_xH_y are anthropogenic pollutants and are present in all flue gases.

It is important to mention the significant role played by the Sun in regard to the heat (temperature) balance of Earth. The mass of the Sun is almost 95% of the total solar system. The Sun is mainly composed of hydrogen gas (92%), plus about 8% helium and other elements (Manuel, 2009).

The heat reaching the Earth from the Sun is very dynamic. The daily cycle (i.e., sunrise/sunset) imparts dynamic heat transfer. Further, Sun surface is found to show flare activity, which is variable over time. The latter is also found to be variable. Further, the clouds in the Earth's atmosphere reflect the heat from sun. Since there are different kinds of clouds, the degree of reflection of the Sun's rays is dependent on this latter parameter. All in all, one thus finds that heat transfer from the Sun to the Earth is highly variable from day to day (as well as in time and space). In addition, one must mention that the difference in temperature on Earth from day to night can be as high as 20°C. Further, the solar energy reaching the Earth consists of different wavelengths (as seen from diffraction through a prism). The range of light from Sun is from 380 nm (violet) to 750 nm (red):

VIOLET, INDIGO, BLUE, GREEN, YELLOW, ORANGE, RED

(WAVELENGTH: $nm = 10^{-9} m$)

The mechanism of climate change on Earth would thus be expected to be rather a complex phenomenon, reflected in average global temperature change. For many decades, it was suggested that the heat (infrared) from the Sun is absorbed by GHGs, and this is expected to induce an increase in the temperature of the Earth. GHGs (such as CO_2, CH_4) are molecules in air that absorb infrared light. This leads to heat absorption from the reflected Sun's energy falling on the Earth (Li et al., 2011). It is estimated that the average power density of solar radiation at the outside of the atmosphere of Earth is approximately 1366 W/m², which is called the solar constant. Further, it is found from this figure and the surface area of the Earth that the annual solar energy reaching the Earth is ca. 5,460,000 EJ/year. It is important to compare this with the annual energy consumption by man, which is about 500 EJ. This means that mankind consumes/emits only 0.01% of solar energy.

CHANGE OF CONCENTRATION OF CO_2 IN AIR (AND CO_2 IN OCEANS)

It is also important to mention the possible reason/source for the CO_2 increase in air in the recent century. With the rapid increase of the global population and the industrialization of more and more countries, the demand for energy is growing almost proportional to the world population. Currently over 85% of global energy demand is supplied by the burning (combustion) of fossil fuels (e.g., natural gas + oil + coal). It is also expected that fossil fuels will continue to be the main source of energy in the foreseeable future. The burning (combustion) of these fossil fuels releases CO_2 into the atmosphere; this disturbs the carbon (CO_2) balance of the planet, which has been steady over hundreds of millions of years. Although anthropogenic CO_2 emissions are relatively small as compared to natural carbon fluxes, such as photosynthetic fluxes, the increased release has had obvious influences on the global climate in a very short period of time (Robinson et al., 2007; IPCC, 2007, 2019).

The estimates of CO_2 as found in different states (either emissions or as stored) are estimated as follows (as Gt of carbon, C):

Anthropogenic CO_2 from fossil fuel combustion (variable): 8 Gt C/year

CO_2 as present in biosphere and oceans: 40,000 Gt C

CO_2 in atmosphere: 780 Gt C

It is obvious from these data that oceans are an important factor as regards CO_2 (material balance) emission/capture technology. CO_2 found in the oceans is considered as being a buffer, which can absorb/release any amount of CO_2 in air as needed (Appendix C). Analyses show that since the beginning of the industrial age in ca. 1750, the CO_2 concentration (average) in the atmosphere has increased from 280 to 410 ppm in 2017 (Figure 1.2) (IPCC, 2007, 2019).

The increase of the CO_2 concentration in atmosphere influences the balance of incoming and outgoing energy (i.e., heat) in the atmosphere system (owing to its GHG properties), leading to the rise of the average surface temperature of Earth. Thus, CO_2 has often been cited as the primary anthropogenic GHG, while other GHGs are also expected to contribute to this phenomenon.

In different industries where fossil fuels are burned (through the process of combustion), the flue gases are known to contain some other pollutant gases (such as CO, NO_x, SO_2, etc.) besides CO_2. In all cases, flue gases from coal/oil-fired plants have been purged of these different pollutants for many decades. However, CO_2 from these flue gases has not been captured to the same extent.

The timescale of the data for CO_2 (Figure 1.2) is obviously short as compared to geological timescales. However, there are studies in the literature that do give (indirect) estimates of Earth data for much longer timescales (Appendix C).

It is also found that the seasonal variation of CO_2 is approximately ±5 ppm (i.e., summer/winter). The concentration of CO_2 drops in summer owing to the uptake through *photosynthesis* in plants. This thus indicates that approximately 5 ppm of CO_2 is captured by plants/forests, a variation of ca. ±10% during the summer months. CO_2 is found to be soluble in water (Appendix C). The solubility increases (ca. three-fold) at 0°C as compared to at 30°C. In other words, in oceans (where one finds a very large amount of CO_2), the concentration of CO_2 would vary with temperature (i.e., from summer to winter) (IPCC, 2007).

Geological studies have shown that prior to the modern industrial human technology, the surface and atmosphere of the Earth maintained a steady state with a relatively constant amount of CO_2, about 280 ppm, for several millennia before the invention of fire (and steam engines, etc.) (Appendix C). This may also indicate that the **minimum** concentration of CO_2 in air would be approximately 280 ppm. Furthermore, equilibrium between the two states (i.e., CO_2 in air and CO_2 in oceans) (equation 1.1) must be maintained to sustain the necessary growth of food/plants and to support life. Over hundreds of millennia, the atmospheric CO_2 concentration (and Earth's temperature) has varied naturally, owing to variations in the Earth's tilt and its orbit around the sun. The natural state of the climate for the past million years or more has been cold (glacial ice ages) with periodic warm (interglacial) periods. It is also reported that the average global temperature has increased by about 1°C since the Industrial Revolution (Appendix C).

1.1.1 Temperature of the Earth (Sun—Atmosphere—Earth)

The average temperature of the Earth is primarily controlled by the Sun's radiation. The temperature of the Earth that is discussed in the literature is measured at some fixed points on the Earth (Appendix C).

SUN—ATMOSPHERE—EARTH

This subject is beyond the scope of this book, but a short mention is useful for the general discussion of climate change. It is obvious that the amount of heat the Earth receives from the Sun is a rather complex phenomenon. The Earth is also known to both receive and lose heat through different processes. It is thus useful to consider briefly the comparative quantity of sources of heat involved in the Earth's climate.

All the planets in the solar system are entirely controlled by the Sun (for example, as regards different properties of the solar planets such as temperature, path, etc.).

Measurements show that the average power density of solar radiation just outside the atmosphere of the Earth is 1366 W/m² (solar constant) (Chen, 2011). It is however found that the temperature of the Sun (surface temperature of the sun: 5500°C) is increasing owing to chemical reactions. Further, the energy reaching the Earth is fluctuating (both as a function of time and place). The solar activity (flares and other eruptions) vary daily and yearly.

From this data one has estimated the energy falling on Earth from Sun in the following process (Manuel, 2009; Chen, 2011; Sheer, 2013):

Total energy of solar radiation reaching
Earth per year = 5.46 × 10²⁴ J = 5,460,000 EJ/year.

It is found that this quantity is very large as compared to some anthropogenic energy sources. The annual global energy consumption (between the years 2005 and 2010) has been reported to be about 500 EJ. This amounts to ca. 0.01% (500/546,000 × 100) of the annual solar energy reaching Earth. In these calculations, the average solar power on the Earth was assumed to be 1366 W/m². However, not all solar radiation that falls on Earth's atmosphere reaches the ground. About 30% of solar radiation is reflected into space. About 20% of solar radiation is absorbed by clouds and gas molecules in the air. About three-quarters of the surface of the Earth is covered by water (oceans). The oceans therefore absorb a large part of the heat from sunshine. This phenomenon is further complicated by the fact that water shows some abnormal behavior as regards density and temperatures. However, theoretically, even if only 10% of total solar radiation is utilizable, 0.1% of it can power the entire world.

1.2 CARBON CAPTURE AND STORAGE (CCS) (SURFACE CHEMISTRY ASPECTS)

After the capture of CO_2 by the CCS technology, the carbon (CO_2) needs to be stored or used in some useful application (Chapter 4). This step in CCS has been investigated, and many studies have been reported. The economic aspects of CCS have been also investigated (Rackley, 2010).

The **purpose of this book** is to describe the surface chemistry aspects of carbon capture (i.e., CO_2) (Dennis, 2014). The storage step may consist of different processes, depending on the nature of the application. The CCS technology thus in simple terms means (Figure 1.4):

STEP I: Anthropogenic CO_2: fossil fuel combustion (burning)
STEP II: Anthropogenic CO_2 ==== CO_2 capture and storage (CCS)
STEP III: Storage methods for captured CO_2

It is important to specify the major sources of CO_2 (where CO_2 is present in **free state**):

a. *atmospheric CO_2;*
b. *anthropogenic CO_2 (from fossil fuel burning);*
c. *CO_2 dissolved in oceans/lakes.*

This means that **free** CO_2 is in equilibrium in the I and III states (Appendix C). In other words, if its concentration changes in one state, it will change in the other states correspondingly (e.g., increase or decrease). State II is variable. The CCS technology can thus be applied to these two (I and II) major states where CO_2 is present in the free state. Further, since CO_2 is soluble in water (1.45 gm/liter), large quantities are thus present in oceans/lakes/rivers (Appendix C).

Furthermore, since CO_2 in the atmosphere is essential for the growth of plants on Earth, there must be a **minimum** of its concentration in order to maintain life on Earth.

CO_2 Equilibrium:

Minimum Concentration of CO_2 in Air === CO_2 in Oceans

One may estimate this minimum CO_2 concentration (without any anthropogenic CO_2 emissions) to be equal to approximately 280 ppm. This figure is related to the average CO_2 concentration before the beginning of the Industrial Revolution. CCS technology thus has to address the following parameters (Ionel et al., 2008; Metz et al., 2005; IEA, 2014b; Beising, 2007; Scheer, 2007; IPCC, 2007; Hinkov et al., 2016; Leung et al., 2014; Rackley, 2010):

control/capture of anthropogenic contributions of CO_2 to air;

reduce/control the CO_2 in atmosphere.

The main CCS approach relates to:

Anthropogenic CO_2 production (combustion of fossil fuels)

Increase in CO_2 concentration in air

Capture of CO_2 (at the source: flue gas, directly from air)

CCS thus relates to control and monitoring of CO_2 concentration in air. After this capture process, there is a need for storage or similar technology. A short specific survey of CO_2 capture techniques will be given to the reader to help in future research trends and analyses (Chapter 5; Appendix B). Most of the carbon capture processes currently are mainly related to surface chemical principles (adsorption, absorption, membrane separation, cryogenic process, CO_2 hydrate formation, etc.) (Section 1.2.1). Currently, the main research on CCS is based on two main surface chemical processes: adsorption and absorption. This is the main theme of this book. The surface chemistry aspects will be delineated with the help of classical thermodynamics of adsorption theory (as described by Gibbs adsorption theory) (Chattoraj & Birdi, 1984; Adamson & Gast, 1997; Bolis, 2013; Yang, 2003; Myers & Monson, 2002; Keller et al., 1992; Yu et al., 2012; Birdi, 1997, 1999, 2003, 2009, 2010, 2014, 2016, 2017; Rackley, 2010). However, some additional carbon capture techniques will also be briefly mentioned.

The concentration of CO_2 as found in different (major) sources (natural and anthropogenic) varies as follows:

in atmosphere (ca. 400 ppm = 0.04%)

flue gas from a coal/oil-fired plants (ca. 10%)

flue gas from a natural gas (>95% CH_4) fired plant (ca. 2%)

This variation means that the CCS technology needed for these sources will be different (as related to the concentration of CO_2). Further considerations as regards the parameters of CO_2 variables:

CO_2 concentration in air has increased from 280 ppm to 400 ppm over a few centuries;

CO_2 is soluble in water (e.g., oceans, lakes, rivers) (CO_2 concentration has also increased in oceans).

This shows that anthropogenic CO_2 leads to an increase in its concentration both in air and in oceans/lakes/rivers (due to CO_2 solubility in water).

It is estimated that the anthropogenic CO_2 from fossil fuel combustion is about 25 million tons/year (increased by about 15% over the past 50 years). The cost of application of CCS technology has been evaluated

(Appendix C). This subject is out of the scope of this book and will be discussed only briefly.

It is thus seen that to define the average concentration of CO_2 in the atmosphere is not quite simple. The general process of capture of CO_2 (CCS) is (Figure 1.1; Appendix C):

CO_2 (as gas) ==== CO_2 (capture process: absorbed/adsorbed/etc.)

The most plausible process generally suggested is to capture CO_2 at its production site (i.e., flue gas from coal-, oil-, or gas-fired plants). Currently there are various procedures that are suggested to capture CO_2 from the flue gas of industrial plants. At present, there are two well-established procedures of capturing gases from flue gases (Figure. 1.3):

absorption (Section 2.2)

(gas is captured by interaction with a suitable fluid)

adsorption (Section 2.1)

(gas is captured by interaction with a solid)

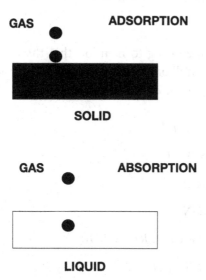

FIGURE 1.3   Gas adsorption and absorption processes.

Other procedures for carbon capture (Rackley, 2010) (Chapter 5; Appendix C).

Binding of CO_2 to minerals (oxides, hydroxides) to form carbonates

Polymer membrane gas separators (size exclusion)

In the case of gas absorption in fluids, the gas molecules have to

1. pass (as bubbles) through the surface of the fluid;
2. diffuse into the fluid (under stirring).

This means that the overall process will include various steps (Chapters 2 and 5). CO_2 may dissolve in the fluid, or it may interact with some component(s) in the solution.

AMOUNTS OF CO_2 IN DIFFERENT STATES

It is important to consider the **quantities** of CO_2 involved for CCS technology. The global CO_2 emissions from different fossil fuel combustion processes are reported to be as follows (approximate data in gigatons (Pg = Gt) CO_2/year):

GAS (5 Pg) + OIL (10 Pg) + COAL (9 Pg) = TOTAL CO_2 EMISSIONS
= 24 Pg

As reference, it is interesting to mention that the current consumption of oil globally is 100 million barrels per day. The various sources of CO_2 (wherever fossil fuels are used) emissions are

CHEMICAL INDUSTRY 1

CEMENT INDUSTRY 1

STEEL INDUSTRY 2

TRANSPORTATION 5

ELECTRIC POWER GENERATION 10

OTHER INDUSTRIES (ETC.) 5

(1 Pg = 10^{15} g = 1 gigaton (Gt) = 10^9 metric tonnes = 10^{12} kg)

1.2.1 Mechanisms of Adsorption at the Gas—Solid Interface

The gas molecules interact with the surface molecules of the solid (when in close proximity). The degree of interaction (i.e., amount of gas adsorbed per gram of solid) depends on various parameters (Chapter 2):

surface forces between gas and solid molecules

temperature/pressure

Experiments show that the gas—solid interaction consists of different kinds of surface forces (Adamson & Gast, 1997; Defay et al., 1966; Chattoraj & Birdi, 1984; Kubek et al., 2002; IPCC, 2007; Radosz et al., 2008; Yang et al., 2017; Shafeeyan et al., 2014; Birdi, 2010, 2014, 2017; Hinkov et al., 2016).

The adsorption energy is mainly dependent on the distance between molecules.

The simple description of gas adsorption on a solid surface can be described as follows. It is known from experiments that the surface molecules in a solid, S, are different than the molecules/atoms inside the bulk phase. A clean solid surface consists of molecules that are the same as in its bulk phase (**S**):

SYSTEM: SOLID (under vacuum)

SSSSSSSSSSSSSSSSSSSSSSSS	surface layer of solid
SSSSSSSSSSSSSSSSSSSSSSSS	bulk phase of solid
SSSSSSSSSSSSSSSSSSSSSSSS	bulk phase of solid

If a gas, G, is present, then it may adsorb with varying degree:

SYSTEM: GAS + SOLID

(MONO-LAYER ADSORPTION) (Figure 1.4A)

GGGGGGGGGGGGGGGGGG	gas adsorption (monolayer)
SSSSSSSSSSSSSSSSSSSSSSSS	surface layer of solid
SSSSSSSSSSSSSSSSSSSSSSSS	bulk phase of solid
SSSSSSSSSSSSSSSSSSSSSSSS	bulk phase of solid

Or it may show bi-layer gas adsorption (Figure 1.4B) (adsorbed gas molecule = **G**):

GGGGGGGGGGGGGGGGGG

GGGGGGGGGGGGGGGGGG   high gas adsorption (bi-layer)

SSSSSSSSSSSSSSSSSSSSSSS   surface layer of solid

SSSSSSSSSSSSSSSSSSSSSSS   bulk phase of solid

SSSSSSSSSSSSSSSSSSSSSSS   bulk phase of solid

Or multilayer adsorption (Figure 1.4B):

GGGGGGGGGGGGGGGGGG

GGGGGGGGGGGGGGGGGG

GGGGGGGGGGGGGGGGGG   extensive high gas adsorption (multi-layer)

SSSSSSSSSSSSSSSSSSSSSSS   surface layer of solid

SSSSSSSSSSSSSSSSSSSSSSS   bulk phase of solid

SSSSSSSSSSSSSSSSSSSSSSS   bulk phase of solid

This schematic drawing merely shows the various surface forces that are involved in the gas adsorption on a solid process (e.g., interactions between gas (G) and solid (S): G—G; G—S). The mechanism of gas adsorption can be determined from the experimental data. Experiments show that these models are indeed found in everyday technology. The most significant feature is that the distance between the gas molecules is reduced by a factor of 10 after adsorption. The adsorption data (i.e., amount of gas adsorbed/gram of solid, pressure of gas) allows one to analyze these mechanisms (Chapters 2 and 5).

1.2.2 Nature of Solid Surfaces (Used for Gas Adsorption)

The nature of the solid (adsorbent) has a profound effect on the gas adsorption process. The basic feature of a good adsorbent (solid) is a large specific surface area (area/gram). For example, the surface area of a solid material/gram increases as the radius of the particles decrease:

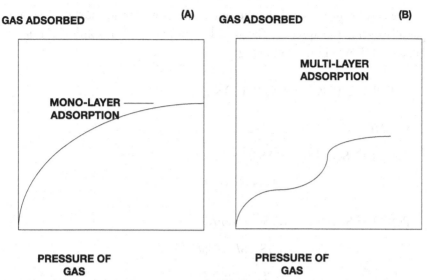

FIGURE 1.4   Gas—solid adsorption isotherms ((A) mono-layer, (B) multi-layer systems) (amount of gas adsorbed versus gas pressure).

Surface area (m²/gm) of solid   Increases with decrease of particle size

Example: Large solid sample particle

Weight = 1 gm   *Volume of solid = 1 cc*
Shape of the solid = cube
1 gm solid with surface area = 1 cm² × 6 = 6 cm²

After crushing to make cubes of size 0.10 cm:

Number of particles (in powder) = 1/(0.1³) = 1000 particles / gm
Surface area of powder (1 gm) = 1000 × 6 × 0.1² = 60 cm²
Increase in surface area / gram = 60/6 = 10 times

It is thus seen that as the number of particles per gram increases, the total surface area of the solid (per gram) increases. Typical solids from everyday life:

Talcum   10 m²/gm

Active charcoal   >1000 m²/gm

In the case of **porous solids,** the pores add considerable surface area for gas adsorption (Chapter 2). The gas adsorption in porous solids is found to behave as shown in the following:

BEFORE GAS (G) ADSORPTION:

PLANE

CRYSTAL SOLID   SSSSSSSSSSS

                   SSSSSSSSSSS

POROUS SOLID   S//S/ /S/ /S/ /S/ /

                   S/ /S/ /S/ /S/ /S/ /

                   S/ /S/ /S/ /S/ /S/ /

(here // indicates pores)

AFTER GAS (G) ADSORPTION:

PLANE   GGGGGGGG

CRYSTAL SOLID   SSSSSSSSSSS

                   SSSSSSSSSSS

POROUS SOLID   **GGGGGGGGGGGG**

                   S/**G**/S/**G**/S/**G**/S/**G**/S/**G**/S/

                   S/**G**/S/**G**/S/**G**/S/**G**/S/**G**/S/

The larger the surface area of the solid adsorbent (per gram of solid), the more gas molecules can be adsorbed on its surface (i.e., per gram of adsorbent). This means that the process is efficient as regards rate, cost, etc. However, the gas adsorption process in porous materials is complex, since it relates to the dimensions of the pores and as well as to the size of the gas. It is also found that the adsorption energy is different inside the pores as compared to outside the pores. Generally, this means that a good adsorbent is very porous. The characteristics of porous solids have been investigated extensively in literature (Appendix C) (Keller et al., 1992; Yang, 1987; Birdi, 2014, 2017). This arises from the fact that pores create extra surface for adsorption. The magnitude can vary from 100 m² to >1000 m²/gm (Keller et al., 1992; Korotcenkov, 2013; Birdi, 2003, 2016).

Porous adsorbents generally may contain pores ranging from *micropores* with pore diameters of less than 1 nm to *macropores* with diameters of >50 nm. These absorbents are found to exhibit properties of the solid surface and range from crystalline materials like zeolites to highly disordered materials such as activated carbons. The typical adsorbents that are used for gas adsorption are (Appendix B; Chapter 5)

silica gel,

activated carbon,

zeolites,

metal organic frameworks,

ordered mesoporous materials,

and carbon nanotubes.

The specific area of an adsorbent is the surface area available for adsorption per gram of the adsorbent. Special procedures (gas adsorption experiments) are used to determine this quantity for a given solid. This quantity is a specific physical property of the solid surface. In order to understand the gas adsorption process, this quantity is needed. Since the gas molecules (**G**) adsorb on the surface:

GAS PHASE

G G G G

————— SURFACE OF SOLID

In order to estimate the area per molecule of the adsorbed gas, one needs to determine:

amount of gas adsorbed,

surface area of solid material per gram.

If one knows the gas molecule dimensions (from other measurements), then one can estimate the packing of the adsorbed layer(s) (Chapter 2). This means that one needs the following data:

amount of gas adsorbed/surface area of solid/gram

area/molecule of gas

It is also obvious that in all gas—solid systems, one needs the information about the surface area per gram of the adsorbent, in order to make physical analyses. The mechanism of adsorption also requires the information about the amount of gas adsorbed per gram of solid as sorbent (as a function of temperature and pressure) (Figure 1.5) (Chapter 5).

1.2.3 Gas—Solid Adsorption Isotherms

One may expect that the process of gas adsorption on a solid may be relatively simple. However, experiments show that it is a somewhat complex system (Chapters 2 and 5). The gas adsorption process is generally investigated by measuring the amount of gas adsorbed per surface area of solid material. The amount of gas adsorbed on a solid is measured as a function of pressure (Figure 1.5). The effect of temperature is also measured in most instances.

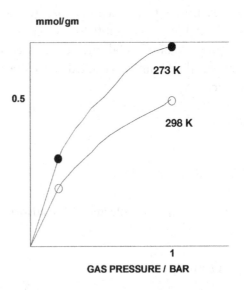

FIGURE 1.5 Gas adsorption on solid (effect of temperature): amount of gas adsorbed (mmol/gm) versus gas pressure.

In general,

gas adsorption on solid surface increases with higher pressures;

gas adsorption on solid surface increases at lower temperature.

The following is CO_2 adsorption data on activated carbon (Suzuki, 1990):

PRESSURE	0.2 bar	1.0 bar
273 K	0.3 mmol/gm	0.7 mmol/gm
298 K	0.2 mmol/gm	0.5 mmol/gm

Similar kind of adsorption behavior is found for various other gas—solid adsorption systems (Chapters 2 and 5). A literature review of CO_2 adsorption on solids is given in Chapter 5.

1.2.4 Gas Absorption in Fluids Technology

This process is described as the gas absorption in a fluid (Figure 1.3). The gas molecule may be soluble or interact with the components in the fluid. The removal of a gas from flue gas by bubbling (scrubber) through a fluid has been used for many decades (Rochelle, 2009; Cents et al., 2001).

The solubility characteristics of CO_2 in water are extensively investigated in the literature (Appendix C). It is reported that CO_2 solubility is 900 cc per 1000 cc of water at STP (standard temperature and pressure: 20°C; 1 atmosphere). However, CO_2 dissolves in water to form carbonic acid, H_2CO_3. The latter dissociates into $H^+ + HCO_3^-$. The equilibria of these species in water has been investigated (Appendix C).

For example, aqueous solutions (with monoethanolamine) have been used to capture CO_2 from flue gases (Chapter 5). In this CCS technology, aqueous solutions of various amines (with basic properties, e.g., monoethanolamines) have been used. These solutions of amines preferably strongly bind CO_2 at low temperatures. However, at higher temperatures it is found that CO_2 is desorbed from the solution. The reaction is depicted as follows:

$$CO_2 + 2\ HOCH_2CH_2NH_2 \leftrightarrow HOCH_2CH_2NH_3^+ + HOCH_2CH_2NHCO_2^-$$

The recovered CO_2 is around 90% pure. This can then be stored or used in different applications (Chapter 4; Appendix C). The gas absorption method for carbon capture has been studied extensively in the current literature. A short literature survey is given later in the text (Chapter 5).

1.2.5 Carbon Dioxide Captured by Different Processes

Besides adsorption and absorption CCS technologies of capturing CO_2, there are also other phenomena that can capture CO_2. For example:

Plants (of all kinds) grow through a process called photosynthesis. Photosynthesis captures CO_2 from air (Paoletti et al., 2002). It has been suggested that increased plantation of forests/trees thus will assist capture of CO_2.

Formation of inorganic salts from CO_2: calcium carbonate, magnesium carbonate, etc.

Capturing CO_2 from a source such as any industrial plant (flue gas) or directly from air (Dubey et al., 2002).

CO_2 is soluble in water (Chapter 4). Hence, a very large quantity of CO_2 is present in oceans/lakes/rivers. At normal temperature and pressure, solubility of CO_2 is about 1.45 gm/liter.

SOLUBILITY OF CO_2 IN WATER (OCEANS/LAKES/RIVERS)

Another very significant chemical process is the equilibrium between CO_2 in air and CO_2 in oceans. This arises from the fact that

about 75% of the Earth is covered by oceans (plus lakes and rivers);

CO_2 is soluble in water (1.45 gm/liter).

The state of the CO_2 equilibrium between air and oceans/lakes/rivers is different than one finds between air and land. Further, this property, that is, the equilibrium between the CO_2 in air and water, is instantaneous. The chemical equilibrium between the concentration of CO_2 in the oceans and atmosphere can be described as follows (Figure 1.5):

$$\mu_{CO_2,\,air} = \mu_{CO_2,\,ocean} \tag{1.1}$$

The equilibrium constant, K_{CO_2}, is

$$K_{CO_2} = C_{CO_2,\,air} \,/\, C_{CO_2,\,ocean.} \tag{1.2}$$

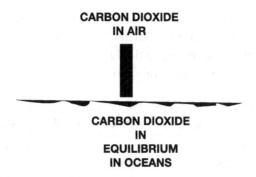

**CARBON DIOXIDE
IN AIR**

**CARBON DIOXIDE
IN
EQUILIBRIUM
IN OCEANS**

FIGURE 1.6   Carbon dioxide equilibrium between air <> oceans/lakes/rivers.

where $C_{CO_2,air}$ and $C_{CO_2,ocean}$ are concentrations in air and ocean, respectively (Appendix C), and $\mu_{co_2,air}$ and $\mu_{co_2,ocean}$ are the chemical potentials of CO_2 in air (atmosphere) and ocean, respectively (Appendix C). This means that if the CO_2 concentration changes in either of the phases, the equilibrium changes its concentration in the other phase. Thus, if the concentration of CO_2 increases in air, then it will induce an increase in concentration in ocean/lakes (as follows from the chemical potential equilibrium equation 1.1). In other words, currently there are 410 ppm CO_2 in air, which is in equilibrium with the CO_2 in the oceans. If the concentration of CO_2 is reduced (or changes) to 300 ppm in the air, then some equivalent amount of CO_2 will be released from the oceans, to satisfy the chemical equilibrium state (equations 1.1 and 1.2). Most significant is that this equilibrium is instantaneous, and the amounts involved (in the oceans) are also very large in comparison to anthropogenic CO_2 from fossil fuel combustion (Dennis, 2014). However, CO_2 as found dissolved in the oceans (lakes/rivers) does not contribute to the greenhouse phenomena. It is reported that ca. 95 Pg of carbon (CO_2) is transported across the atmosphere—ocean interface. The CO_2 equilibrium at the ocean surfaces is very fast. It is estimated that about one-third of anthropogenic CO_2 is absorbed by the oceans (equation 1.2) (IPCC, 2007). In other words, the oceans can absorb and store considerable amounts of CO_2 (owing to their large volume).

Furthermore, because of the anomalous physical properties of water, the CO_2—water (i.e., oceans) equilibrium is not as simple as one may assume as a first consideration. Water is the only fluid that exhibits a maximum density (at 4°C) (Birdi, 2014, 2017). This gives rise to some exceptional features. For instance, as temperature reaches 4°C, the water at the surface flows to the bottom of ocean. This means the circulation in oceans due to temperatures around 4°C will also contribute to the CO_2 cycle since it is soluble in water. The tidal phenomena in oceans is related to the gravity interaction with the moon. These tidal phenomena give rise to turbulence and mixing in the upper layers of oceans.

1.3 APPLICATION OF CCS ON CONTROL OF CARBON DIOXIDE CONCENTRATION IN AIR

It has been asserted by the Intergovernmental Panel on Climate Change (IPCC) that CO_2 emissions (anthropogenic) could be reduced by 80–90% for a modern power plant that is equipped with suitable CCS technologies (Figure 1.1). The application of CCS technology will also complement other crucial strategies, such as changing to fuels with less carbon content, improving energy efficiency, and phasing in the use of renewable energy resources (Phelps et al., 2015; Rackley, 2010). In a different context, atomic energy/hydro-energy/solar energy/wind energy are almost CO_2-free technologies.

Direct Capture of CO_2 From Air

Direct air capture refers to the process of removing CO_2 directly from the ambient air (as opposed to from point sources). As mentioned elsewhere, there is a minimum concentration of CO_2, ca. 280 ppm, and hence the current CO_2 concentration is ca. 120 ppm in excess (400 – 280 ppm). It is obvious that because of the very low concentration of CO_2 in air, methods must be modified accordingly.

Geological Storage of CO_2

After CO_2 is captured (generally about 90% pure), it needs to be stored in some safe process (or used in some suitable application). This technology is about handling of captured CO_2. The captured CO_2 has been stored according to various procedures. This procedure has been also called *geosequestration*; this method involves injecting CO_2, generally in liquid state, into suitable underground geological formations (Appendix C):

Old coal/oil reservoir

other storage structures

It is expected that various physical (e.g., highly impermeable caprock) and geochemical trapping mechanisms would prevent the CO_2 from escaping to the surface (Birdi, 2017).

Oil recovery from reservoirs is a multistep process. Most of the recovery is primarily based on the original pressure in the reservoir (producing around 20% of oil in place (Chapter 4) (Birdi, 2017). CO_2 has been used in oil recovery processes; it is sometimes injected into declining oil fields to increase oil recovery. It is found that about a few hundred million metric tonnes of CO_2 are annually injected worldwide into these oil reservoirs.

1.3.1 Flue Gases and Pollution Control

The content of CO_2 in flue gases is different in various technologies. For example, flue gas from coal-fired plants contains 9 to 14% CO_2 (besides other gases). The natural gas-fired (mostly methane, CH_4) plants have flue gas with ca. 4% CO_2 content.

It is estimated that without the application of CO_2 control (such as CCS technology or other) technology, the CO_2 concentration in air would increase to 650–1550 ppm by 2030 (Li et al., 2011; 2019; Phelps et al., 2015). It is also generally suggested that because different pollutants from flue gas are captured/removed, CO_2 should also be treated the same way. In a different context, the technology of cleaning flue gas (of different pollutant gases: CO, NO_x, SO_2) is very advanced at the present stage. Therefore, CCS is suggested as a viable technology (Chapter 5). Of course, the quantities involved are much different, since the concentration of CO_2 is much higher than that of the other gas pollutants (Appendix C).

CHAPTER 1 SUMMARY

The current concentration of CO_2 present in air is ca. 400 ppm. The concentration has increased slowly since the Industrial Revolution (1750). The slow increase (ca. 2 ppm/year) is reported to be related to fossil fuels (e.g., coal, oil, natural gas (methane) burning; Chapter 4). In the past decades, CO_2 from flue gases or air has been captured by use of specific technologies. These are called carbon capture and sequestration (CCS) processes (Chapter 4). These are based on different CCS technologies (e.g., absorption/adsorption, CO_2 hydrate, CO_2 freezing, membrane separation, etc.) for CO_2 gas. In the absorption process the gas is absorbed in alkaline fluids (aqueous solutions; Chapter 2). In the adsorption press the gas is adsorbed on a suitable solid. In most of these processes, surface chemistry principles are of primary interest. The latter aspects of CCS are described in the following text (Chapters 2 and 5).

Adsorption (on Solids) and Absorption (in Fluids) of Gases (CCS Procedures) (Surface Chemistry Aspects)

2.1 INTRODUCTION

The concentration of CO_2 in air (atmosphere) has been found to be increasing over the past century (Figure 1). This is generally attributed to increasing use of the fossil fuel combustion process in technologies. Owing to its GHG characteristics, it is thus obvious that mankind should develop and use technology that can control the increase of CO_2 from this anthropogenic process. Since the Industrial Revolution, mankind has increasingly added CO_2 to the atmosphere from fossil fuel combustion technology.

Carbon capture and storage (CCS) technology is currently based on removal by various processes (Figure 1.1):

absorption in liquid (fluid) (by passing gas through fluid (scrubber));

adsorption on solids (by passing gas through solid); other various techniques

(membrane gas separation, CO_2 hydrate formation, cryogenic method)

In this chapter some of the major processes used for capturing any gas will be described. The CO_2 emission control from different industrial (power) plants (flue gases) requires the removal of this GHG from the flue gas. In the current literature, extensive studies are reported on the application of absorption/adsorption processes (Bishnoi & Rochelle, 2000; Myers & Monson, 2002; Rochelle, 2009; Harlick & Tezel, 2004; Yu et al., 2012; Rackley, 2010). Similar approaches (with modifications) can be applied to removal/extraction of CO_2 directly from air (Appendix C).

Surface chemistry principles are applied in all these two-phase (or multi-phase) processes (e.g., where a substance A adsorbs/absorbs on another material, B). In the case of gas (A_{gas})—solid or gas—fluid system:

$$\text{GAS MOLECULES}\left(A_{gas}\right) + \text{SOLID/FLUID MOLECULES}(B) = \text{ADSORBED/ABSORBED}\left(A_{gas}\right) + \text{SOLID/FLUID}(B) \tag{2.1}$$

In the case of a gas—solid system, the adsorbed gas molecules are then desorbed by changing the experimental conditions (such as temperature/pressure). In the case of a gas—fluid system, the absorbed gas is desorbed/purged from the fluid by changing the experimental conditions. Both **adsorption** (Section 2.2) and **absorption** (Section 2.3) processes lead to enrichment of the material being captured. For example: the adsorption/absorption process is a method of obtaining enriched gas (>90% CO_2) from a mixture (as from a flue gas (5–10% CO_2). This may have positive economic consequences in the end use of the recovered CO_2. Pure CO_2 is used in various applications in everyday use (such as in the food industry, cleaning technology, oil recovery, etc.). This observation is important since the extra expense needed for CCS can be compensated to some degree by such applications.

These two different major gas-capturing methods will be described in this chapter. Furthermore, some additional information about the subject in this chapter will be explained (Appendix B).

2.2 ADSORPTION OF GAS ON SOLIDS

The interaction of gases with solids has been investigated for over a century (Langmuir, 1917; Chattoraj & Birdi, 1984; Adamson & Gast, 1997; Yang, 1987; Bolis et al., 2008; Keller et al., 1992; Myers & Monson, 2002; Birdi, 2003, 2017; Hinkov et al., 2016). Adsorption of gases on solid surfaces has been investigated in much detail in the literature. This arises from the fact that this phenomenon is of importance in many everyday technology processes and natural phenomena. One of the most common observations is the corrosion of metals (such as iron, zinc) when exposed to air. In the corrosion process, oxygen from air interacts with the metal (base metal) to produce metal-oxide. Oxygen is also known to interact with different materials on Earth, and oxidation chemical processes are very important in different cases (both useful and not useful). In fact, corrosion is a very undesirable process and is one of the costliest processes for mankind. Another extensively studied phenomena is catalyst technology. Catalyst technology is also an important surface chemical application technology. One can remove pollutants from flue gases by adsorbing/absorbing the toxic gases such as N_2O_x, SO_2. In other words, in everyday processes, one has these two kinds of gas—solid adsorption processes.

The gas adsorption on solids thus comprises two parts (Figure 2.1):

(INITIAL STAGE)

GAS PHASE   G   G   G   G (ADSORBENT)

SOLID PHASE   SSSSSSSSSSSSSS (ADSORBATE)

(FINAL STAGE)

gas molecules (**adsorbent**)   GGGGGGGGGGGGG

solid (adsorbate)   SSSSSSSSSSSSSSS

It is thus seen that the gas molecules are in a different state after adsorption than in the original gas phase. The surface atoms of a solid interact with the environment (such as gas molecules (in the present context), liquids, etc.) to a varying degree because of surface forces. **Adsorption** is the process whereby molecules from the gas (or liquid) phase are taken up by a solid surface; it is distinguished from **absorption,** which refers to

**GAS ADSORPTION
PROCESS**

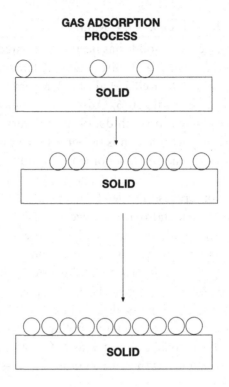

FIGURE 2.1   Adsorption stages of gas on a solid.

molecules entering into the lattice (bulk) of the solid material or a liquid phase.

The **adsorptive** is the material in the gas phase capable of being adsorbed, whereas the **adsorbate** is the material actually adsorbed by the solid. The solid, which exposes the surface sites responsible for the process, is called the **adsorbent**. In Figure. 2.1 the adsorption process at the surface of a solid material is schematically illustrated.

The concentration of gas will thus be high near the solid surface, due to adsorption (Figure 2.2). This shows that the adsorption leads to a process where the adsorbed gas is present at a very high concentration, as compared to its original state. This is the most significant effect of surface adsorption process. The process of all kinds of gas adsorption on solids is governed by either *physical* or *chemical* forces. In the former case the adsorption is named physical adsorption (*physisorption*); in the latter case,

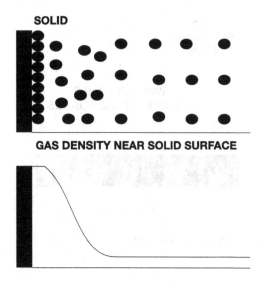

FIGURE 2.2 Gas adsorption on a solid: concentration profile of gas near the solid surface (formation of mono-layer of gas).

chemical adsorption (*chemisorption*). The different mechanisms and the surface forces involved will be described (Appendix A).

The interaction between gas molecule/atom and solid is determined by the interaction energy (energies). When two entities (bodies) come in close (i.e., atomic dimension) proximity, there are present different kinds of interactions. Some of the most common kind of interactions energies are as follows (Adamson & Gast, 1997; Chattoraj & Birdi, 1984; Bolis et al., 2008; Birdi, 2003; 2016, 2017; Hinkov et al., 2016) (Figure 2.3):

dispersion forces (short range);

electrostatic (long-range forces);

and chemical bonds.

In the case of physical adsorption, the adsorbate (gas)—adsorbent (solid) potential of adsorption will be dependent on various forces

$$\pi = \pi_D + \pi_I + \pi_E \tag{2.2}$$

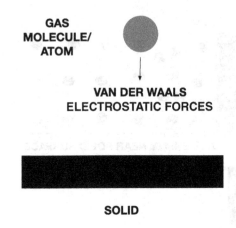

FIGURE 2.3   Gas—solid interactions (different surface forces).

π_D = dispersion energy,

π_I = induction energy (interaction between electric field and an induced dipole),

π_E = interaction between electric field (F) and a permanent dipole (μ),

which are operative in all adsorbate—adsorbent systems. The π_E potential interaction contribution arises from charges (on the gas and solid surface) (Masel, 1996; Birdi, 2016). However, in the case of chemical interactions (where a chemical bond is formed) the process is very different. For instance, the oxidation of metals (such as iron, aluminum) by oxygen (as found in air), leads to the formation of a new compound (e.g., Fe_2O_3, Al_2O_3). Of course, when the concentration of gas is very low (as in the case of CO_2 in air), chemical bond formation is preferable to other techniques.

For example, in the case of activated carbon, the nonspecific interactions dominate. For metal oxides, zeolites, and ionic solids, the electrostatic interactions often dominate, depending on the adsorbate (Barrer, 1978; Masel, 1996; Razmus & Hall, 1991; Gregg & Sing, 1982; Steele, 1974; Adamson & Gast, 1997; Rigby et al., 1986; Young & Crowell, 1962; Ross, 1971; Birdi, 2003). The interaction potentials are related between an atom/molecule (or a charge) on the surface and the adsorbate atom/molecule. All interactions between two bodies depend primarily on the magnitude of the distances separating these, R_d. The expression for the π_D is:

Dispersion forces

$$\pi_D = - A_D R_d^{-6} + B_D R_d^{-12} \qquad (2.3)$$

where A_D and B_D are constants. The expression for πI (field of an ion) and induced point dipole:

$$\pi_I = -2 \alpha F \qquad (2.4)$$

Gas adsorption on a solid is a process analogous to the separation of molecules (such as chromatography). The energetic of the process is thus dependent on the gas phase and solid phase characteristics. The degree of interaction between the molecules in gas phase and solid thus can be investigated. The distance between molecules/atoms in different phases can be estimated as described below.

In the case of all-natural processes, one defines the different phases of matter:

Gas

Liquid

Solid

Further, when any two or more phases meet, one has a **surface** (or **interface**) to consider (Chapter 3; Appendix A).

SOLID PHASE/INTERFACE/GAS PHASE

SOLID PHASE/INTERFACE/LIQUID PHASE

SOLID (1) PHASE/INTERPHACE/SOLID (2) PHASE

And in the case of liquids:

*LIQUID PHASE/**INTERFACE**/GAS PHASE*

*LIQUID (1) PHASE/**INTERFACE**/LIQUID (2) PHASE*

Surface chemistry principles (Chapter 3 and Appendix A) apply to the state of properties of these different "interfaces." The state of the matter (phase) at any given temperature or pressure is given by the distance between the molecules and the energy of interactions between the molecules.

SOLID PHASE: The molecules in solid state are closely bound to each other, and molecular forces keep the shape of a solid.

LIQUID PHASE: In the liquid state, the molecular forces between molecules are somewhat weaker than in solids. Thus, a liquid takes the shape of its container.

In general, the distance between molecules in solid phase is shorter than in the liquid phase (ca. 10%). However, water molecules behave differently. Ice is lighter (ca. 10%) than liquid water (Appendix A). That is why ice floats on water (for example, icebergs float on water).

In order to explain this in more semi-quantitative detail, let us consider water. Water is found in liquid state at room temperature and pressure (1 atmosphere). In order to estimate the distance between the molecules, the following data is useful (standard state):

1 mole of water (18 gm) (H_2O) (at room temperature and pressure)

Volume of one mole (water) = 18 cc

Volume of 1 mole of gas (all gases) = 22.4 × 1000 cc = 22.4 liters = 22,400 cc

Ratio of volume of gas:liquid = 22,400/18 = ca. 1000

This analysis shows that in general, a molecule in the gas phase occupies **1000** times more volume than in the liquid or solid phase. This means that the molecules in the gas phase move larger distances (ca. 10 times) than in the liquid or solid phase. Further, in general, in the solid phase the molecule occupies ca. 10% less volume than in the liquid phase. However, water is an exception, which shows that ice is ca. 10% lighter than water at 0° (hence ice floats on water). The latter observation has other consequences, as regards absorption of gases in ice (Appendix C). Consequently, this also means that oceans/lakes have additional physical properties as regards absorption and diffusion of gases (especially CO_2 which is soluble in water) (Appendix C).

In general, **adsorption** is a process where one substance (atoms or molecules) of one substance in one phase (in the present case gas/fluid phase) interacts (physical interaction or chemical interaction) with the surface of a different phase (surface of a solid).

2.2.1 Theory of Adsorption of Gas on Solid Surfaces (Basic Remarks)

Gas molecules are in continuous movement and possess kinetic energy. Furthermore, the gas molecules experience surface forces when they come in proximity to a solid (surface). Experiments show that the gas molecule (or atom) may adsorb on the solid surface with varying surface forces. The degree of gas adsorption is found to be determined by solid surface forces. (Chapter 3).

At temperatures above the critical point, there is a continuous change from the liquid to the vapor state (based on van der Waals theory: at the boundary between a liquid and its vapor there is not an abrupt change from one state to the other, but rather a transition layer exists in which the density and other properties vary gradually from those of the liquid to those of the vapor) (Langmuir, 1918; Adamson & Gast, 1997; Myers & Monson, 2002; Chattoraj & Birdi, 1984; Keller et al., 1992; Birdi, 2003). Later, this postulate of the continuous transition between phases of matter has been applied generally in the development of theories of surface phenomena, such as surface tension, adsorption, absorption, catalysis, foams, bubbles, etc. (Adamson & Gast, 1997; Keller et al., 1992; Birdi, 2003).

Further, similar to the theory described to account for surface tension phenomena. one assumes that the molecules in the transition layer are attracted toward one another with a force that is an inverse exponential function of the distance between them. Accordingly, experiments show that there is an abrupt change in physical properties in passing through the surface of any solid or liquid. The atoms/molecules that are in the surface of a solid are held to the underlying atoms by forces similar to those acting between the atoms inside the bulk of the solid phase. From crystal structure studies and from many other considerations, it is known that these forces are of the type that have usually been designated as chemical. In the surface layer, because of the asymmetry of the conditions, the arrangement of the atoms must always be slightly different from that in the interior of the bulk phase. It is seen that these atoms will be unsaturated chemically and thus they are surrounded by an intense field of force.

It has been suggested that when gas molecules impinge against any solid or liquid surface they do not in general rebound elastically but may adsorb on the surface, because they are under the field of force of the surface atoms. Additionally, gas molecules lose kinetic energy after desorption.

These gas (adsorbed) molecules may subsequently desorb from the surface of the solid (Figure 2.3):

GAS MOLECULES → MOVE (IN THREE DIMENSIONS) WITH KINETIC ENERGY

GAS MOLECULE → IMPINGES ON SOLID

GAS MOLECULE BOUNCES BACK ⇔ NO ADSORPTION

GAS MOLECULE ADSORBS ⇔ ADSORPTION

The length of time that elapses between the adsorption of a molecule and its subsequent desorption will depend on the degree of intermolecular surface forces (Figure 2.4). Further, in the case that the surface forces are relatively strong, desorption will be observed at a very low rate, which will lead to the surface of the solid becoming completely covered with a layer of molecules. In the case of gas adsorption, this layer will usually be not more than one molecule deep, for as soon as the surface becomes covered by a single layer (monolayer) the surface forces are chemically saturated. However, if the surface forces are weak, the desorption may take place. This means after the adsorption, only a small fraction of the surface becomes covered by a single layer of adsorbed gas molecules. In accordance with the chemical nature of the surface forces, the range of these forces has been found to be extremely small, of the order of 10^{-8} cm. That is, the effective range of the surface forces is usually much less than the diameter

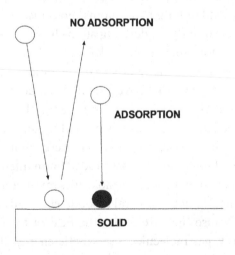

FIGURE 2.4   Gas—solid adsorption mechanisms (see text).

of the gas molecules. The gas molecules thus usually orient themselves in definite ways in the surface layer since they are bound to the surface by forces acting between the surface and particular atoms or groups of atoms in the adsorbed molecule. The orientation of adsorbed molecules on the solid surface is thus another specific characteristic (as found from atomic microscope studies) (Birdi, 2002). The dipole of the gas molecule will also determine the absorption interactions energy.

The diameter of molecules of ordinary permanent gases averages about 3×10^{-8} cm (3 Å). There are thus about 10^{15} molecules per cm^2 in a monomolecular layer, corresponding to 0.04 mm^3 of gas per cm^2, at ordinary temperature and pressure; therefore, with surfaces of about a square meter, the amount of gas required to cover the surface with a single layer of molecules is 400 mm^3 (Figure 2.1).

The model (ideal) for gas adsorption on solids is shown as follows:

GGGGGGGGGGGGGGGGGGGGGG   adsorbed gas molecules

SSSSSSSSSSSSSSSSSSSSSSSSSSSSS   surface molecules of solid

SSSSSSSSSSSSSSSSSSSSSSSSSSSSS   bulk solid molecules

SSSSSSSSSSSSSSSSSSSSSSSSSSSSS

It is thus seen that

-the distance between G molecules (i.e., between **G—G**) in the adsorbed state is almost 10 times smaller than in the gas state;

-the distance between G and S (i.e., **G—S**) is also reduced after adsorption.

Extensive gas adsorption studies on different solid surfaces (for example, mica, glass, platinum, zeolites, carbon, coal, etc.) have been carried out (Chapter 5).

For example, in the case of solids such as platinum, gas adsorption has been found to provide useful information of a metallic surface; where the surface is not covered with a layer of oxide or other surface film that might absorb gases, it was necessary to use a non-oxidizing metal such as platinum.

The adsorption potential will be dependent on the surface forces at the interface: gas molecule/atom—solid.

The interaction energy will thus be different for different systems. Therefore, in any industrial application one needs to know these interaction potentials to be able to achieve maximum output.

The surface of a solid can be characterized (as regards the molecular topography) by the following different properties:

plane surfaces,

rough surfaces,

porous solids (Figure 2.5).

The **plane** faces of a crystal are found to consist of atoms forming a regular plane lattice structure. The atoms in the cleavage surface of crystals like mica are those that have the weakest fields of surface forces of any of the atoms in the crystal. It is probable that in mica the hydrogen atoms cover most, if not all, of the surface, since hydrogen atoms when chemically saturated by such elements as oxygen possess only weak residual valence. In the case of solid surfaces such as glass and other oxygen compounds (e.g., quartz, glass, calcite, zeolites), the surface probably consists

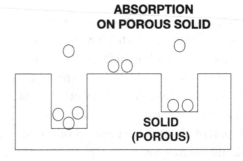

FIGURE 2.5   States of solid surfaces (plane, rough, porous).

of a lattice of oxygen atoms. The surface of crystals thus resembles to some extent a well-defined surface for gas adsorption sites. When molecules of gas are adsorbed by such a surface these molecules take up definite positions with respect to the surface lattice and thus tend to form a new lattice above the old. The new structures of such adsorbed phase can be studied by suitable spectroscopic analyses (Chapter 5).

Hence, a unit area of any crystal surface has a definite number of sites with each selectively able to adsorb a gas molecule or atom. In general, these specific binding sites will not all be expected to be alike, that is, as regards the adsorption energy. It is found that there will frequently be cases where there are two or three different kinds of sites. For example, in a mica crystal it may be that both oxygen (O) and hydrogen (H) atoms, arranged in a regular lattice, form the surface in such a way that different elementary spaces are surrounded by different numbers or arrangement of atoms.

For example, one can consider a (schematic) most plausible arrangement of oxygen (O) and hydrogen (H) atoms in a surface lattice (for example, mica) (Langmuir, 1918; Chattoraj & Birdi, 1984; Adamson & Gast, 1997):

H..O..H..H..O..H

H..H..O..H..H..O

O..H..H..O..H..H

H..O..H..H..O..H

H..H..O..H..H..O

O..H..H..O..H..H

This is a very simplified example of real systems. If the adsorbed molecules are present at sites over the centers of the squared spaces, they tend to form a new lattice above the old. Experiments show that all solids exhibit very characteristic gas adsorption behavior. It is found that there the unit area of any solid surface has a well-defined number of "adsorption sites." Each adsorbing site has a well-defined number of gas molecules/atoms. In general, it is reasonable to expect that not all of the active binding sites will be exactly alike. This may arise from defects in the surface morphology (which has been corroborated from analyses, such as X-ray diffraction, scanning atomic microscopes, etc.).

There are also some solids with surface characteristics where there are two or three different kinds of adsorption sites. For example, in a mica crystal it may be that both oxygen and hydrogen atoms, arranged in a regular lattice, form the surface in such a way that different elementary spaces are surrounded by different numbers or arrangements of atoms. In fact, this kind of surface will be expected in all solid minerals. For example, it has been suggested that one may expect the arrangement of oxygen and hydrogen atoms in a surface lattice as indicated:

HOHHOH HHOHHO OHHOHH HOHHOH HHOHHO OHHOHH

If the adsorbed molecules take up positions (based on geometrical considerations) over the centers of the square shape site, there are two kinds of elementary situations, those represented by **H H** and those represented by **O H**. For each of the latter there are two of the former kind of space. In case the adsorbed atoms/molecules arrange themselves directly above the surface atoms, one will expect that there may be two kinds of gas adsorption sites. These analyses thus suggest from considerations of this kind that a crystal surface may have sites of only one kind or may have two, three, or more different kinds of sites representing definite simple fractions of the surface. Further, each kind of site space will, in general, have a different tendency to adsorb gases. As the pressure of gas is increased, the adsorption will then tend to take place in steps, and the different kinds of spaces will be successively filled by the adsorbed molecules.

One may thus safely conclude that the adsorption of a gas on a solid may show many different types of mechanisms. This is indeed also observed from experimental data (Chapter 5).

METHODS USED FOR GAS ADSORPTION STUDIES

In general, in a system where a solid is exposed in a closed space to a gas at pressure p, the weight of the solid typically increases (due to gas adsorption: as shown in Figure 2.6) and the pressure of the gas decreases as the gas is adsorbed by the solid. At equilibrium, the magnitude of pressure p does not change, and the weight reaches an equilibrium value. A schematic of a simple adsorption setup is given in Figure 2.6. Many commercial apparatuses that are used for measuring the gas adsorption (at varying temperature and pressure) are available (Chapter 5).

The amount of gas adsorbed is experimentally determined (a) by gravimetry (for example: the increase in weight of the solid is monitored

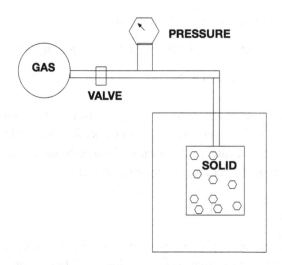

FIGURE 2.6   Schematics of apparatus used for gas adsorption on solid.

by a spring balance); (b) by volumetry (the fall in the gas pressure is monitored by manometers/transducer gauges); (c) by monitoring the change of any other physical parameter related to the adsorption of matter, such as the evolved heat (if the heat of adsorption is known and constant) or the integrated IR absorbance (if the specific molar absorbance of adsorbed species is known). A simple adsorption process of gas molecules on a plane surface having only one kind of elementary site and in which each site can only adsorb one gas molecule/atom can be described as follows.

GAS ADSORPTION ON SOLID (KINETIC MODEL)

Molecules in the gas phase are in continuous motion (as follows from the kinetic theory of gases) (Adamson & Gast, 1997; Birdi, 2003). It is thus obvious that all processes involving gases would be dynamic phenomena. Gas molecules will thus continuously strike the surface of a solid. Depending on the experimental conditions (pressure, temperature), any of the following may happen (Figure 2.6):

gas molecules bounce back to the gas phase (weak or no adsorption);

gas molecule may adsorb;

gas molecules may adsorb and desorb continuously.

DYNAMICS OF GAS—SOLID SURFACE:

GAS MOLECULES CONTINUOUSLY STRIKE SOLID SURFACE:

DEGREE OF ADSORPTION

From this dynamic description of the gas adsorption process, one can derive the relation between the amount of gas adsorbed and the pressure and temperature. The rate at which gas molecules come into contact with a surface is given by the relation for kinetic theory:

$$N_{gas} = \left(M_{gas} \big/ \left(2\,\pi\,R\,T\right)\right)^{0.5} P_{gas} \tag{2.5}$$

Here N_{gas} is the number of grams of gas striking the surface per sq. cm. per second, M_{gas} is the molecular weight, T the absolute temperature, p_{gas} the pressure in bars, and R the gas constant, 83.2×10^6 ergs per degree. One can define a quantity g_{gm}, the number of gram molecules of gas striking each sq. cm. per second, then $g_{gm} = N_{gas} / M_{gas}$:

$$g_{gm} = P_{gas} \big/ \left(2\,\pi\,M_{gas} R\,T\right)^{0.5} \tag{2.6}$$

$$= 43.75\ 10^{-6}\ p \big/ \left(M_{gas} T\right)^{0.5} \tag{2.7}$$

This relation is a modification of the equation giving the rate of effusion of gases passing through small openings (Adamson & Gast, 1997; Birdi, 2003, 2016).

The system of interest is thus:

INITIAL STATE

gas phase (gas molecules G)

solid phase (solid molecules S)

AFTER GAS ADSORPTION:

GAS PHASE (GAS MOLECULES)

GAS ADSORBED

ON SOLID **GGGGGGGGGGGGGGGG**

SSSSSSSSSSSSSSSSSSSSSS

The molecules in the gas phase can be analyzed by the classical kinetic gas theory (Adamson & Gast, 1997; Chattoraj & Birdi, 1984; Myers & Monson, 2002; Sing et al., 2005). Gas molecules move about in space and possess kinetic energy and occupy 1000 times more volume than in the liquid or solid state. It is thus obvious that the state of adsorption on a solid is where the gas molecules are present with comparatively very low kinetic energy.

One assumes a system where the molecule/atom M is adsorbed molecularly (i.e., without rupture/formation of chemical bonds) from the gas at a surface site, S. It is also assumed that the gas molecules only adsorb on specific sites (N_S) on the solid (Chapter 3). Furthermore, all solids have a definite maximum number of sites on the surface (per gram of solid), where gas molecules can adsorb. This number can be estimated by an adsorption method (Bolis, 1998).

The fractional monolayer coverage (θ) of the sites occupied by adsorbate molecules is defined as follows:

$$\theta = N_s / N_{total} \tag{2.8}$$

where N_{total} is the total number of absorbing sites on the solid. The quantity N_s is a unique characteristic of solid. The rate of adsorption is given as

$$\text{adsorption rate} = k_{ads\,Pgas}(1-\theta) \tag{2.9}$$

k_{ads} being the rate constant for the adsorption and $(1 - \theta)$ the fractional monolayer coverage of sites not occupied yet by the adsorbate molecules.

The rate of gas desorption, k_{des}, as related to the rate constant for desorption, is given by the equilibrium (dynamic conditions):

$$(\text{adsorption rate}) = (\text{desorption rate}) \tag{2.10}$$

From this one gets (the Langmuir equation):

$$\text{desorption rate} = k_{des}\theta \tag{2.11}$$

$$k_{ads}\, p_{gas}(1-\theta) = k_{des}\,\theta \tag{2.12}$$

$$\theta / (1-\theta) = K\, p_{gas} \tag{2.13}$$

where K = the ratio of the rate constant for adsorption/the rate constant for desorption, $= k_{ads}/k_{des}$.

The gas adsorption data is analyzed (Langmuir equation: Chattoraj & Birdi, 1984; Birdi, 2003) by the relation:

$$\theta = V/V_{mon} = (K_{P_{gas}})/(1 + K_{P_{gas}}) \tag{2.14}$$

The term V represents the adsorbate volume and V_{mon} the monolayer volume, that is, the volume of adsorbate required to complete the monolayer.

At very low pressure the equation reduces to a linear dependence of the coverage upon the equilibrium pressure ($\theta = hp$). Conversely, at high pressure the equation reduces to the case of coverage approaching the monolayer ($\theta \approx 1$).

The quantity monolayer coverage (V_{mon}), the quantitative magnitude of this is not easily determined experimentally with high accuracy. Hence, by using a different procedure, the Langmuir equation is suitably transformed in the so-called reciprocal linear form:

$$(1/V) = 1/(K\,V_{mon})(1/p_{gas}) + 1/V_{mon} \tag{2.15}$$

In the case where the data fits the Langmuir model, the plot of the reciprocal volume, $1/V$, against reciprocal pressure, $1/p$, is linear (Langmuir-type isotherms). In other words, such data plots provide useful information as regards the gas adsorption equilibrium.

However, if the experimental data plot deviates from this linear relation, this indicates that the Langmuir equation does not agree with the given adsorption process. This observation is analogous to the ideal gas equation. The deviation of adsorption isotherm thus indicates a non-ideal system.

The magnitude of the quantity monolayer capacity is estimated from the intercept $i = 1\,/\,V_{mon}$ of the straight line. From this one can estimate the equilibrium constant K from the slope (s):

$$s = 1/(K\,V_{mon}) \tag{2.16}$$

The monolayer volume and the equilibrium constant are typical of the adsorbent/adsorbate pairs at a given temperature. In particular, the value of K is related to the strength of the adsorbent—adsorbate interaction: high values of K indicate large strength, and low values indicate weak adsorbing forces.

As an example: the data of a typical system, that is, case of CO adsorbed at $T = 303$ K on dehydrated Na– and K–MFI will be discussed (as regards the experimental volumetric and calorimetric isotherms). The number of

CO molecules adsorbed per gram of zeolite at p_{CO} represents the number of occupied sites (N_S), whereas the number of charge-balancing cations exposed per gram of zeolite represents the total available sites (N).

From a data plot of the coverage $\theta = N_S/N_{total}$ and the $\theta / (1 - \theta)$ quantity are plotted against p_{CO}. The slope of the $\theta / (1 - \theta)$ versus p_{CO} plot is the Langmuir constant K. The CO molecules are found to exhibit soft Lewis base properties (i.e., they get polarized by the electrostatic field generated by the alkaline-metal cations located in the MFI zeolite nano-cavities). This leads to the adsorption of gas molecules (reversibly) on the surface when in contact with the (activated) zeolite.

The equilibrium constant K for Na—MFI ($4.88 \pm 0.02\ 10^{-3}$ Torr^{-1}) was found to be larger than for K—MFI ($1.15 \pm 0.02\ 10^{-3}$ Torr^{-1}). This was found to be in agreement with the different polarizing characteristics of the cations (e.g., Na$^+$ and K$^+$). In fact, the local electric field generated by the unsaturated cations depends on the charge/ionic radius ratio, which is larger for Na$^+$ than for K$^+$ (ionic radius of Na$^+$ = 0.97 Å and of K$^+$ = 1.33 Å. As regards the charge/ionic radius ratio, the maximum coverage attained at $p_{CO} = 90$ Torr was larger for Na—MFI ($\theta \approx 0.3$) than for K—MFI ($\theta \approx 0.1$). This shows that there is a correlation between adsorption and the size of the cation. For example: The analyses of the adsorption data for CO further showed that the magnitude of (standard free energy) $\Delta G_{ads}°$ for CO adsorption at the two alkaline-metal sites was obtained from the Langmuir equilibrium constant K:

$$\Delta G_{ads}° = -RT (\ln K) \tag{2.17}$$

In both cases the adsorption process in standard conditions was found to be endothermic:

$$\Delta G_{ads}° = +13.4 \text{ kJ mol}^{-1} \text{ for Na—MFI;}$$

$$\text{and } \Delta G_{ads}° +17.0 \text{ kJ mol}^{-1} \text{ for K—MFI.}$$

Further studies also showed that after vacuum treatment the two Na$^+$ $\cdots$ CO and K$^+$ $\cdots$ CO adspecies were absent. This shows that CO was absent after the vacuum treatment of the solid.

From $\Delta G_{ads}°$ the standard entropy of adsorption, $\Delta S_{ads}°$, can be estimated if $\Delta H_{ads}°$ is known. The CO adsorption enthalpy change was measured calorimetrically during the same experiments in which the adsorbed amounts were measured.

Generally, the gas adsorption data for real systems shows deviation from the ideal Langmuir adsorption model (especially at high gas pressures) (Chapter 5). According to the assumptions of Langmuir theory:

(a) the solid surface is rarely uniform: there are always "imperfections" at the surface;

(b) the mechanism of adsorption is not the same for the first molecules as for the last to adsorb. When two or more kinds of sites characterized by different adsorption energies are present at the surface (as stated in point a), and when lateral interactions among adsorbed species occur (as stated in point b), the equivalence/ independence of adsorption sites assumption fails. The most energetic sites are expected to be occupied first, and the adsorption enthalpy ΔH_a (per site), instead of keeping a constant, coverage-independent value, exhibits a declining trend as long as the coverage θ increases.

Further, one may also find that in some systems on the top of the monolayer of gas, other molecules may adsorb and multi-layers may build up (BET) model (Adamson & Gast, 1997; Birdi, 2003). However, in the case of the adsorption at surfaces characterized by a heterogeneous distribution of active sites, the results were found to be different. Furthermore, one has also found that the gas adsorption data on a solid (the Freundlich equation) may exhibit an isotherm that will fit the following expression:

$$V_{ads} = k_{p_{gas}}{}^{1/n} \qquad (2.18)$$

This relation is based on simple empirical considerations, where the term V_{ads} represents the adsorbed amount and p_{gas} the adsorptive pressure, whereas k and n are empirical constants for a given adsorbent—adsorbate pair at temperature T.

In some gas adsorption isotherms, the data is found to fit the following equation (an exponential equation: Temkin isotherm):

$$V_{ads} = k_1 \ln(k_2\, p) \qquad (2.19)$$

It is known that this relation is a purely empirical formula, where V_{ads} represents the adsorbed amount and p the adsorptive pressure; k_1 and k_2 are suitable empirical constants for a given adsorbent—adsorbate pair (at temperature T). The isotherm data suggests that the adsorption enthalpy ΔH_{ads} (per site) decreases linearly upon increasing adsorption.

Examples of heats of adsorption decreasing linearly with coverage are reported in the literature, as for instance in the case of NH_3 adsorbed on hydroxylated silica, either crystalline or amorphous. Similar results were obtained in the case of CH_3OH adsorption on silica-based materials.

Further, it is worth noticing that at sufficiently low pressure, all-gas adsorption isotherms are linear and may be regarded as obeying **Henry's law** (Figure 2.7):

$$V_{ads} = h_{H\,Pgas} \qquad\qquad (2.20)$$

Henry's law of gas adsorption relates the amount of gas adsorbed to the pressure, p_{gas}. The Henry constant h_H is typical of the individual adsorbate—adsorbent pair and is obtained from the slope of the straight line representing the isotherm at low adsorption coverage. The isotherms classification, which is of high merit in terms of generality, deals with ideal cases that in practical work are rarely encountered. In fact, most often the adsorption process over the whole interval of pressure is described by an experimental isotherm that does not fit into the classification.

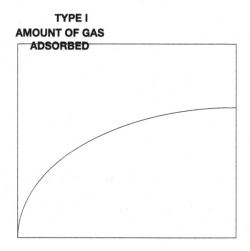

FIGURE 2.7A   Gas—solid adsorption isotherms (amount of gas adsorbed versus gas pressure; Type I).

TYPE II
AMOUNT OF GAS
ADSORBED

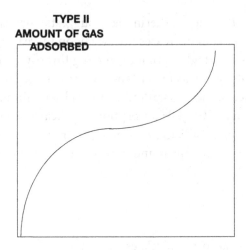

RELATIVE GAS PRESSURE

FIGURE 2.7B   Gas—solid adsorption isotherms (amount of gas adsorbed versus gas pressure; Type II).

TYPE III
AMOUNT OF GAS
ADSORBED

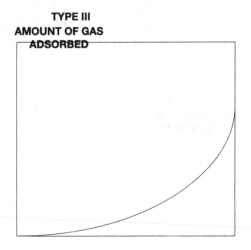

RELATIVE GAS PRESSURE

FIGURE 2.7C   Gas—solid adsorption isotherms (amount of gas adsorbed versus gas pressure; type III).

Nonetheless, each of the equations described above may be applied over a given range of equilibrium pressures. This procedure thus allows one to analyze the experimental isotherm through the combination of individual components to the process. By using this approach, the surface properties of the solid and the thermodynamics features of processes taking place at the interface can be quantitatively analyzed (Attard & Barnes, 1998; Adamson & Gast, 1997).

For example, the adsorption data of NH_3 on a dehydrated silica specimen were found to be consistent with the Langmuir and Henry isotherm equations. This behavior was explained to be due to the presence of hydrogen bonding (H bonding) on silanol (Si-OH) groups.

NUMBER OF ADSORBING SITES (FOR GAS MOLECULES) ON A SOLID

All solid surfaces will be expected to have a specific **maximum** number of adsorption sites. This is also found from experiments. One may denote N_0 as the number of elementary sites per cm^2 of surface. Then, the number of gas molecules adsorbed cannot exceed N_0 except by the formation of additional layers (multi-layers) of molecules. The forces acting between two layers of gas molecules will usually (expected) be very much less than those between the solid surface and the first layer of molecules. In any case, two cases that are different, for example, G—S or G—G, will be expected to be under different interactions. This means that the energy of interaction is different:

GGGGGGGGGGGGGGG

SSSSSSSSSSSSSSSSSS

SSSSSSSSSSSSSSSSSS

SSSSSSSSSSSSSSSSSS

The double layer:

GGGGGGGGGGGGGGG

GGGGGGGGGGGGGGG

SSSSSSSSSSSSSSSSSSS

SSSSSSSSSSSSSSSSSSSS

MOLECULAR THEORY OF GAS ADSORPTION ON SOLIDS

In any adsorption phenomenon the substances from the external environment (e.g., gas or liquid) are absorbed by a solid surface (adsorbent). The adsorption process has been used to separate gaseous and liquid mixtures, for drying and purification of gases and liquids. The adsorption calculation of the equilibrium and dynamic characteristics of adsorption in porous bodies at the molecular level have been investigated in literature.

In a recent study adsorption theories based on statistical physics were derived to explain the adsorption process (Tovbin, 2017). These were found to be the consistent with the description of the equilibrium distribution of molecules and dynamics of flows in complex porous materials. These data were found to be useful for a wide range of practical applications in the development of new technologies.

Furthermore, the effect of gas pressure on adsorption is found to be as follows:

> at low gas pressure, the amount of adsorbed gas is proportional to the pressure but increases much more slowly at higher pressures. However, if the relative adsorption rates for the two gas species are different from each other, then experiments show that the adsorption isotherms show definite steps as the pressure increases.

2.2.2 Solids with Special Characteristics of Gas Adsorption

In everyday one finds solids which one uses exhibit with different structures. Some of the different solid characteristics are given in the following.

Adsorption on Amorphous Surfaces: In general with crystal surfaces there are probably only a few different kinds of elementary spaces, but with amorphous substances (such as glass, zeolites), the elementary binding sites may all be different. One may thus expect the surface divided into infinitesimal fractions, with each possessing its own binding site.

Furthermore, one finds gas—solid systems with some very unique properties. Experiments show that the surface forces that hold adsorbed substances act primarily on the individual atoms rather than on the molecules. When these forces are sufficiently strong, it may happen that the atoms leaving the surface become paired in a different manner from that in the original state molecules. These examples of gas adsorption are found in different **catalysts**. The synthesis of ammonia (NH_3) from nitrogen and hydrogen (by using a catalyst) is one of the most important examples.

However, the adsorption of a diatomic gas such as oxygen (O_2) or carbon monoxide (CO) on a solid surface needs different analyses. One assumes that the atoms are individually held to the solid surface, each atom occupying one elementary space. The rate of desorption of the molecules is expected to be negligibly small, but occasionally adjacent atoms combine together and thus nearly saturate each other chemically, so that their rate of evaporation becomes much greater. The gas molecules thus leave the surface only in pairs. For example, initially the system consists of a solid surface with no adsorbed gas molecules.

Starting with a bare surface, if a small amount of gas is adsorbed, the adjacent atoms will nearly always be the atoms that adsorb together when a molecule was adsorbed. In some cases, however, two molecules will happen to be adsorbed in adjacent spaces. One atom of one molecule and one of the other may then desorb from the surface as a new molecule, leaving two isolated atoms that cannot combine together as a molecule and are therefore compelled to remain on the surface. At equilibrium, one may expect there will be a haphazard distribution of atoms over the surface.

Furthermore, there may also be systems of gas—solid surface where there will be two kinds of adsorbing sites, one which is occupied by the gas molecules:

GGGGGGGGGGGGGGGG

SSSSSSSSSSSSSSSSSSS

SSSSSSSSSSSSSSSSSSS

SSSSSSSSSSSSSSSSSSS

and one where there will be un-occupied sites (indicated as . . .):

GGG . . . GGG . . . GGG . . . GGG

SSSSSSSSSSSSSSSSSSS

SSSSSSSSSSSSSSSSSSS

SSSSSSSSSSSSSSSSSSS

In order that a given molecule approaching the surface may adsorb (and be retained for an appreciable time) on the surface, two particular elementary sites must be vacant. The gas adsorption thus becomes dependent on the chance of the molecules hitting the two kinds of sites.

Adsorption of More than One Gas Molecule in Thickness (multi-layer adsorption) on Solids: With gases or vapors at pressures much below saturation, the surface of a solid tends to become covered with a single layer of molecules. The reason for this is that the forces holding gas molecules (or atoms) on to the surface of solids are generally much stronger than those acting between one layer of gas molecules and the next. When the vapor becomes nearly saturated, however, the rate of evaporation from the second layer of molecules is comparable with the rate of condensation so that the thickness of the gas **film** may exceed that of a molecule. These thicker films may also be present in those cases where the forces acting between the first and second layers of adsorbed molecules are greater than those holding the first layer to the surface. An example of this latter kind has been found experimentally in the condensation of cadmium vapor on glass surfaces.

2.2.3 Gas Adsorption Isotherms

It is thus seen that the adsorption process at the gas/solid interface is related to an enrichment of one or more components in an interfacial layer (Chattoraj & Birdi, 1984; Sing et al., 1985; Adamson & Gast, 1997; Keller et al., 1992; Birdi, 2003):

GAS MOLECULE (G). G. G. G.

ADSORBED GAS MOLECULE (**G**)

GGGGGGGGGGGGGGGGGGGGG

SOLID SURFACE SOLID SURFACE

The mechanism of gas adsorption (e.g., amount of gas adsorbed per unit gram of solid (with a specific area/gm), number of layers of gas adsorbed) is determined from the analyses of experimental data isotherms. The gas adsorption will be dependent on temperature and pressure for each system. The adsorption mechanisms (for example, monolayer, bi-layer, multilayer) will lead to different kinds of isotherms. This arises from the fact that the energetics of adsorption is dependent on the arrangement of gas molecules. These data from the literature have shown that (Sing et al., 1985) there exist distinct **six different types of adsorption** systems (Figures 2.7 to 2.12). The atomic structures of the adsorbed phase have been also identified in some cases (by using atomic microscopic analyses).

The mechanism of gas adsorption (e.g., mono-layer, bi-layer, or multilayer) on any solid surface can be determined from the experimental data. The **shape of the adsorption isotherm** has been found to provide useful information as regards the physical characteristics of the adsorbate

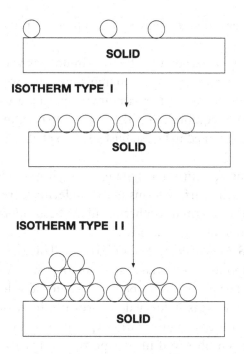

FIGURE 2.8   Gas structures corresponding to different isotherms in Figure 2.7 (Types I and II).

(gas) and solid adsorbent (in the case of porous solids, the pore structures) geometry (Adamson & Gast, 1997; Keller et al., 1992; Birdi, 2003, 2017; Silin & Kneafsey, 2012; Sing et al., 1985).

TYPE I GAS ADSORPTION ISOTHERM: The most commonly applied adsorption model for gas reservoirs (for example, shale reservoirs) (Yu et al., 2014a/b; Birdi, 2016) is the classic Langmuir isotherm (Type I) (Langmuir, 1918), which is based on the assumption that there is a dynamic equilibrium at constant temperature and pressure between adsorbed and non-adsorbed gas. Also, it is assumed that there is only a single layer of gas molecules adsorbed on the solid surface (Fig. 2.7A). Figure 2.8 depicts the adsorbed gas monolayer structure on the surface, which corresponds with the inflection in the isotherm in Figure 2.7A. The monolayer structure is related to the calculation based on the amount of gas adsorbed and the area/molecule data. In the adsorption of this type (generally called the Langmuir isotherm), the isotherm data can be analyzed by the relation

$$v(p) = (v_L \ p)/(p + p_L) \qquad (2.21)$$

where v(p) is the gas volume of adsorption at pressure p and v_L is the Langmuir volume.

The quantity v_L corresponds to the maximum gas volume of adsorption at infinite pressure and p_L is Langmuir pressure, which is the pressure corresponding to one-half Langmuir volume. It is assumed that there exists instantaneous equilibrium of the sorbing surface, and the storage in the pore space is assumed to be established (Gao et al., 1994; Keller et al., 1992).

It is also found that in some systems, at high gas pressures, the gas adsorbed on the solid surface forms **multi-molecular layers** (Figure 2.7). In other words, the Langmuir isotherm will not be an appropriate approximation of the amount of gas adsorbed.

TYPE II GAS ADSORPTION ISOTHERM: This type of adsorption isotherm corresponds to multilayer sorption of gas, and the gas adsorption isotherm of Type II should be expected to be valid (Figure 2.7). After the monolayer gas is formed, as pressure increases, multi-layer structures are formed (Figure 2.8). Experiments show that Type II isotherms are often observed in non-porous or macroporous materials (Kuila, 2013; Freeman et al. theory in the Journal of the American Chemical Society; Brunauer et al., 1938; Sing et al., 1985). The BET isotherm model is a generalization of the Langmuir model to multiple adsorbed layers (as shown in Figure 2.7B). The BET model of adsorption assumes a homogeneous surface of the solid and that there exists no lateral interaction between the adsorbed molecules. The equation of BET also assumes that the uppermost layer is in equilibrium with the gas phase.

The data plot of p/v ($p_o - p$) versus p/p_o (generally a straight line) gives an estimate of the value of $1 / v_m C$ at the intercept and the slope is equal to the quantity:

$$(C-1)/v_m C \tag{2.22}$$

From these data and v_m, the specific surface area can be estimated. The standard BET isotherm theory assumes that the number of adsorption layers is infinite.

TYPE III GAS ADSORPTION ISOTHERM: This type of adsorption isotherm is only observed in some rare adsorbent materials (Figure 2.9). One instance is the adsorption of nitrogen on ice (Adamson & Gast, 1997). There is some indication of multi-layer formation (Figure 2).

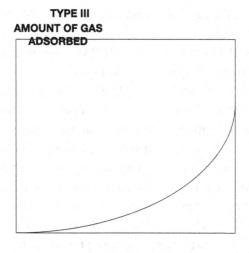

FIGURE 2.9 Gas—solid adsorption isotherms (amount of gas adsorbed versus gas pressure (Type III).

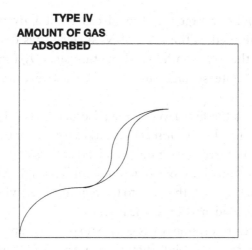

FIGURE 2.10 Gas—solid adsorption (and desorption) isotherms (amount of gas adsorbed versus gas pressure; Type IV).

The adsorption of gas does not form a monolayer; instead, clusters are formed.

TYPE IV GAS ADSORPTION ISOTHERM: This type of isotherm is indicative of mono-layer/bi-layer/multi-layer structures (Figure 2.10). This type is also called the capillary condensation type. This type is typically found in porous solids.

TYPE V GAS ADSORPTION ISOTHERM: These isotherms are generally observed with multi-layer adsorption systems.

TYPE VI ADSORPTION ISOTHERM: The Type VI isotherm (stepped) (Figure 2.13), which is relatively rare, is also reported in literature. Type IV (Figure 2.11) and V isotherms typically exhibit a hysteresis loop, which is characteristic of porous systems, involving capillary condensation.

It is thus seen that the process of gas adsorption on a solid is described through isotherms, that is, through the functions connecting the amount of adsorbate (i.e., gas atom/molecule) taken up by the adsorbent (solid) (or the change of any other physical parameter related to the adsorption of matter) with the adsorptive equilibrium pressure p, the temperature T, and all other parameters being constant. Below the critical temperature the pressure is properly normalized to the saturation vapor pressure p°, and the adsorbed amounts are hence referred to the dimensionless relative pressure p/p_{o}.

The **fractional coverage** θ of the adsorbate, at a given equilibrium pressure p, is defined as the ratio of N_S surface sites occupied by the adsorbate over the total available adsorption sites N_T, that is, the total number of substrate surface sites that are active toward the given adsorptive.

More specific is the **first** layer of the adsorbed phase. The adsorption energy arises from either *chemisorption, physisorption*, or both, according to the nature of the forces governing the adsorbate/adsorbent interactions. Conversely, the **second** layer of the gas molecules is originated by physical forces, similar to the forces that lead to the non-ideal behavior of gases and eventually to the condensation to the liquid.

The **multi-layer** adsorption of gas layers is expected to approach a liquid-like phase structure on the solid. This means gas molecules are packed close and will thus be expected to behave like a liquid or solid (solid-like).

In the adsorption process, as the number of N_S occupied sites approaches the number of total available sites N, the adsorbate monolayer is complete ($\theta = 1$). For any further adsorption of gas, the latter will have to adsorb on gas molecules (which are adsorbed).

In Figure 2.11, the formation of subsequent layers of adsorbate at the surface of a solid sample is schematically illustrated. The process of gas adsorption is dependent on various experimental parameters. The

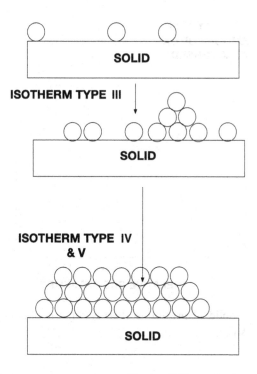

FIGURE 2.11   Gas structures corresponding to different isotherms in Figure 2.7 (Types III, IV, & V).

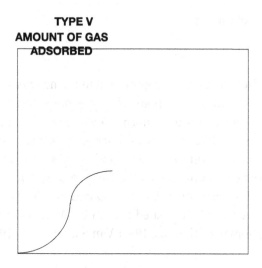

FIGURE 2.12   Gas—solid adsorption isotherms (amount of gas adsorbed versus gas pressure; Type V).

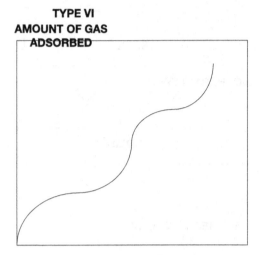

TYPE VI
AMOUNT OF GAS
ADSORBED

RELATIVE GAS PRESSURE

FIGURE 2.13 Gas—solid adsorption isotherms (amount of gas adsorbed versus gas pressure Type VI).

amount of gas taken up by a solid surface depends upon the solid and the gas nature:

the pressure p of the gas,

the temperature T.

The degree of adsorption being proportional to the mass m and the surface area A of the sample, adsorbed amounts (often expressed as mass or volume at the *STP* of the gas) are properly normalized either to the unit mass or to the unit surface area. Here, in view of describing the process at molecular details, the adsorbed amounts n_{ads} are properly expressed as adsorbate moles (or molecules) per either unit mass or unit surface area of the adsorbent.

For example: The adsorption data for CO (at $T = 303$ K) on Na– and K–MFI zeolites, have been investigated (both by volumetric and calorimetric methods) (Thomas & Thomas, 1997; Van Santen et al., 1999; Hench & Wilson, 1993).

2.2.4 Solid Surface Structures

When a molecule (or an atom) from the gas phase approaches a solid, it is more or less strongly attracted by the atoms exposed at the surface, according to the nature of both the gas molecule and the solid material (Chapter 3).

The molecular arrangement of the solid phase has different characteristics based on the single-crystal structure. For example, a crystalline solid is described through the periodic infinite repetition of an elemental pattern (unit cell) [1]. However, the symmetry of the periodic repetition of the unit cell terminates at the surface, thus creating asymmetrical force interaction. This molecular description has been studied in molecular detail by suitable atomic microscopes.

The surface atoms arrangement depends on the plane preferentially exposed during the formation of the surface, according to the partition conditions of the real material (either in the single-crystal form or as nano-sized powder). If no major reconstruction processes are required in order to minimize the surface atoms energy, and if no structural/compositional defects are present, an ideal perfect homogeneous surface is obtained that can be properly represented by cutting a slab of the solid structure. Such an ideal perfect homogeneous surface is very rarely encountered, unless especially prepared for surface science studies.

Real solid surfaces (mostly in the case of finely divided, nano-metric sized solids) are made up of a combination of:

flat regions (terraces): -_-__-_-__-_---__--

structural defects, such as

steps: ---__---

kinks:—

corners: /\\|\||\|\

edges: ---|||||---

point defects: --—-—--

vacancies of ions/atoms in the solid: + +++ +++ --—--.

These surface characteristics thus suggest that gas adsorption will also be dependent on these parameters (as also found from experiments). Compositional defects may contribute to the "imperfections" of the solid surface. These include a variety of oxidation states of the atoms constituting the solid and/or a variety of heteroatoms present either as impurities or especially introduced in order to modify the physico-chemical properties of the surface. During the past few decades, with the advent of high-resolution electron microscopes and other atomic microscopes, it has become possible to image surfaces of solids with high-resolution atomic details (at nano-meter scale) (Zecchina et al., 2001; Birdi, 2002).

For example, structural defects at the surface have been imaged at molecular scale by high-resolution transmission electron microscopy (HR-TEM), as found for monoclinic ZrO_2 nano-crystals, which terminate with structural defects as steps, kinks, edges, and corners (Bolis et al., 2004). The surface defects are found to give rise to valence unsaturation, thus creating a surface that has highly reactive sites for gas adsorption.

Porous Solid Materials: The surface properties of a solid are dependent on various factors. The most important arises from the size of solid particles. Finely divided solids possess not only a geometrical surface, as defined by the different planes exposed by the solid, but also an internal surface due to the primary particles' aggregation. This leads to pores of different sizes according to both the nature of the solid and origin of the surface. Experiments show that these pores may be circular, square, or other shape. The porous solids (Figure 2):

/S/ /S/ /S/ /S/ /S/ /S/ /S

/S/ /S/ /S/ /S/ /S/ /S/ /S

/S/ /S/ /S/ /S/ /S/ /S/ /S

where pores are indicated as //.

The gas (G) molecules will adsorb in the pores as indicated:

/S/G /S/G /S/G /S/G /S/G /S/G /S

/S/ G/S/G /S/G /S/G /S/G /S/G /S

/S/ G/S/G /S/G /S/G /S/G /S/G /S

It is thus obvious that much larger amounts of gas will be expected to adsorb in porous solid than non-porous materials. This process is also called absorption in solids. The size of pores is designated as the average value of the width, w (Gregg & Sing, 1982).

The width, w, gives either the diameter of a cylindrical pore or the distance between the sides of a slit-shaped pore. The smallest pores, with the range of width $w < 20$ Å (2 nm) are called *micropores*. The mesopores are in the range of a width in the 20 Å ≤ w ≤ 500 Å (2 and 50 nm). The largest pores, with width w > 500 Å (50 nm), are called *macropores* (Cambell, 1988; Birdi, 2017).

The shapes of pores will vary in geometric size and shape (e.g., circular, square, triangular, etc.). The capillary forces (Appendix A) in these pores will thus depend both on the diameter but also on the shape. In general, most solid adsorbents exhibit (for example: like charcoal and silico-alumina, etc.) irregular pores with widely variable diameters in a normal shape. Another important criterion for porous is as regards the connectivity of individual pores. The degree of connectivity is sometimes observed in the gas adsorption isotherms. Some other adsorbents, such as zeolites and clay minerals, are entirely micro- or meso-porous, respectively. In other words, the porosity in these materials is found not to be related to the primary particles' aggregation but is an intrinsic structural property of the solid material (Rabo, 1976; Breck, 1974; Birdi, 2016).

2.2.5 Thermodynamics of Gas Adsorption on Solid

In the system gas phase—solid phase, the gas molecules are different as regards the kinetic movement. From the equation relating free energy (ΔG) and entropy (ΔS) of any system:

$$\Delta G = \Delta H - T\Delta S \qquad (2.23)$$

The basic description as regards the thermodynamics of gas adsorption on solid surfaces has been reported in the literature (Chattoraj & Birdi, 1984; Adamson & Gast, 1997; Keller et al., 1992; Birdi, 2003, 2016).

One finds that the quantity enthalpy (ΔH_{ads}) of adsorption can be measured by a suitable calorimeter. The kinetic theory of gases shows that molecules are moving in space. Gas molecules will thus be expected to lose most of the kinetic movement after adsorption on a solid. As mentioned earlier, the gas molecules are thus at lower entropy after adsorption. The adsorption of a gas at a solid surface is found to be an exothermic process (enthalpy of adsorption). This is expected by the thermodynamic condition for a spontaneous process:

$$\Delta G_{ads} = \Delta H_{ads} - T \Delta S_{ads} < 0 \qquad (2.24)$$

In fact, adsorption is necessarily accompanied by a **decrease** in entropy ($\Delta S_{ads} < 0$) in that the degrees of freedom of the molecules in the adsorbed state are lower than in the gaseous state.

These data showed that the value of ΔH_{ads} (the enthalpy change of gas adsorption) is expected to be negative.

2.2.6 Determination of Heat (Enthalpy) of Gas Adsorption on Solids (From Indirect Non-Calorimetric Methods)

The quantity of enthalpy of any system can be measured by various methods. The direct method is where a suitable calorimeter is used (Appendix B). Many commercial calorimeters (micro-calorimeters) are available with varying sensitivity characteristics. The sensitivity of the calorimeter is selected with reference to the system to be investigated (i.e., the magnitude of the change in temperatures of adsorption process) (Appendix B).

Enthalpy of any system can also be estimated from the change in free energy with temperature. The quantity of enthalpy of any system can also be obtained by procedures other than the direct calorimetric, that is, from the change of free energy with temperature. This equation is called the Clausius-Clapeyron equation (Keller et al., 1992; Adamson & Gast, 1997; Birdi, 2016):

$$h_{ads} = R T^2 \left(d \ln p / d T\right) N_s, \; A \tag{2.25}$$

$$= \Delta H_{ads} \tag{2.26}$$

p_1 and p_2 values at T_1 and T_2, for a given constant coverage θ:

$$\ln (p_1 / p_2) = q_{s1}/R \left((1/T_2)-(1/T_1)\right) \tag{2.27}$$

For example: N_2 adsorption on H−ZSM5 were investigated (in the 104−183 K temperature range). The adsorption process was studied by using the change upon adsorption of the absorbance intensity of the ν_{OH} stretching band at 3616 cm^{-1}, as related to the Brønsted acidic site $Si(OH)^+Al^-$. From these data the adsorption enthalpy was found to be:

$\Delta H_{ad^\circ} = -19.7 \pm 0.5$ kJ mol^{-1} (calorimetric heats of adsorption = 19 kJ mol^{-1}) measured at $T = 195$ K for N_2 on H−ZSM5.

For example: Measurements of adsorption of NH_3 on a H−ZSM5 zeolite were reported:

$\Delta H_{ads^\circ} = -128 \pm 5$ kJ mol^{-1} (calorimetric heat of adsorption ≈ 120 kJ mol^{-1}).

These data thus show good agreement between the two methods for determining the enthalpy of gas adsorption on different solids.

2.2.7 Entropy of Gas—Solid Adsorption

The gas molecules occupy 1000 times more volume than the same molecules in the liquid or solid phase. This means that the molecules in the gas phase have higher entropy than when they are in the adsorbed state (on a solid surface). It is thus important to consider the entropy, S_{ads}, of gas—solid adsorption processes. The entropy change ΔS_{ads} related to the gas adsorption in an ideal system case was estimated by using the statistical mechanics (rotor/harmonic oscillator) formula (Keller et al., 1992).

The argon (Ar) gas atoms that adsorb on a solid surface will lose entropy from the gas phase: the translation entropy, S_{tr}, of the solid, which is fixed in the space, is taken as zero, whereas the free Ar atoms, before interacting with the solid surface, possess a translational entropy S_{tr}, which amounts to 150 and 170 J mol^{-1} K^{-1} at $T = 100$ and 298 K, respectively, at $p_{Ar} = 100$ Torr.

It has been suggested that the adsorbed gas atom/molecule (G) will be expected to possess a lesser degree of translational free energy. After adsorption, it will exhibit some vibrational energy (Figure 2.14).

GAS PHASE . . . . . . **G (KINETIC MOVEMENT IN THREE DIMENSIONS**

ADSORBED STATE G (VIBRATIONAL MOVEMENT)

Assuming that the vibrational frequency of the adsorbed atoms is $\approx$ 100 cm^{-1}, then one finds the magnitude of the vibrational entropy is $S_v = 18$ and 43 J mol^{-1} K^{-1} at $p_{Ar} = 100$ Torr and at T = 100 and 298 K, respectively.

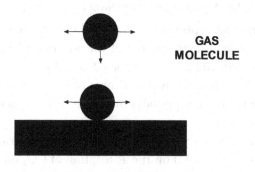

GAS
MOLECULE

SOLID SURFACE

FIGURE 2.14 The kinetic movement of gas molecules in the gas phase, adsorbed on a solid.

As expected, the adsorbed atoms' entropy is much lower than the free gas atoms' entropy, and the entropy change $\Delta_a S = (S_v - S_{tr})$ is in all cases negative: $\Delta S_{ads°} = -132$ Jmol^{-1} K^{-1} at $T = 100$ K and $\Delta S_{ads°} = -127$ J mol^{-1} K^{-1} at $T = 298$ K. This means that in a spontaneous process, which requires a negative free energy change ($-\Delta G$), the enthalpy of adsorption must be negative in order to compensate the loss of entropy. In other words, the process must be exothermic with an amount of heat evolved that is at least as high as the decrease of the T ΔS_{ads} quantity.

The amount of adsorbate depends on the surface strength, and so the negative entropy change (with respect to the free molecules in the gas phase) (Adamson & Gast, 1997; Hill, 1952; Dunne et al., 1997; Garrone et al., 1981; Chattoraj & Birdi, 1984; Birdi, 2003).

The integral molar entropy of adsorption is the difference between the entropy of an adsorbed molecule and the entropy of the adsorptive in the ideal gas state, at given p and T. It is a mean integral quantity taken over the whole amount adsorbed, and it is characteristic of a given state of equilibrium. This is distinguished by the standard integral molar entropy of adsorption, which is the entropy of one adsorbed mole with respect to the entropy of the adsorptive in the ideal gas state at the same T, but under standard pressure.

The quantity of the entropy of adsorption may be obtained from calorimetric experiments only if the heat exchange is reversible. One can estimate the standard adsorption entropy $\Delta S_a°$ from a reversible adsorption volumetric-calorimetric (Otereo et al., 2002; Savitz et al., 1999).

2.2.8 Advanced Experimental Procedures of Gas Adsorption on Solids: STM (Scanning Tunneling Microscope)

The mechanism of adsorption of gas on solid surfaces requires the information about the molecular packing of the adsorbate. This information is useful in the characterizations of the adsorbed gas layer (or layers) in many ways. One needs to know the number of sites/unit surface area of a solid. This information thus provides the data about the packing geometry of the gas on the solid.

In a recent study (Gamba et al., 2016), investigations (using a scanning tunneling microscope, STM) on the adsorption of CO_2 on solid surfaces was reported. As described elsewhere (Birdi, 2002), the molecular details of molecules on surfaces can be investigated by using scanning probe microscopes (SPMs). The atomic scale and microscopic scale structures of surfaces of solids are important data.

The arrangement of the CO_2 molecules as adsorbed on the Fe_3O_4 surface was investigated by STM. These STM images had shown much atomic detail, as regards the adsorption process.

The STM image of the unit cell showed different degrees of greyness depending on the degree of adsorption (Novotny et al., 2013; Parkinson et al., 2011). These data were obtained from STM images following saturation exposure of CO_2 (at a sample temperature of 82 K). The scanning data showed that the position of the bright spots was due to the adsorption of CO_2. Furthermore, it was found that the images of the surface with a submono-layer CO_2 coverage showed *islands*.

2.3 GAS ABSORPTION IN FLUIDS

The solubility of gases in liquids is known to vary. In the case of CCS of CO_2 various fluids have been used. It is known that if a gas is bubbled into a suitable fluid, the latter will dissolve/react to varying degrees. Gas absorption in fluids has been studied extensively, since it is an important phenomenon in everyday life (technical processes and biological phenomena). Some of the most significant examples are the treatment of pollutants removal from flue gases. CO_2 is known to be soluble in water (Chapter 5; Appendix C). Hence, if CO_2 is bubbled in water it will absorb, and is known to form carbonic acid, H_2CO_3.

Solubility of CO_2 in water:

$$\mu_{CO2AIR} = \mu_{CO2WATER} \tag{2.28}$$

The equilibrium constant, $K_{CO2AIR/WATER}$, is:

$$K_{CO2AIR/WATER} = \left(CO_{2WATER}\right)/\left(CO_{2AIR}\right) \tag{2.29}$$

where μ_{CO2AIR} and $\mu_{CO2WATER}$ are the chemical potentials of CO_2 in air and water, respectively;

$K_{CO2AIR/WATER}$ is the equilibrium constant; CO_{2WATER} and CO_{2AIR} are concentrations of CO_2 in water and air, respectively. However, if CO_2 is bubbled in an aqueous solution such as NaOH (sodium hydroxide), then it will react to form Na_2CO_3. This will thus give rise to enhanced absorption. The process of gas capture by absorption in general is described as (Figure 2.10):

GAS BUBBLES—FLUID (SOLVENT)—GAS IN SOLUTION

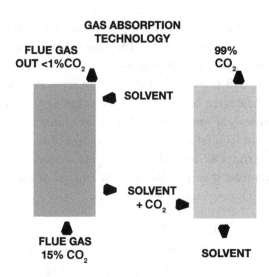

FIGURE 2.15   Gas (CO_2) absorption (scrubber) process.

In most systems the gas interacts with some component in the fluid (solution) (Chiesa & Consonni, 1999; Bishnoi & Rochelle, 2000; Rochelle, 2009; Yu et al., 2012; Hinkov et al., 2016) (Figure 2.15). The process is based on the absorption (selectively) of CO_2 in a solvent (solution). CO_2 is extracted from the second stage of the scrubber (Figure 2.15), where CO_2 (ca. 99%) is recovered. This gives rise to a specific absorption. For example: CO_2 in a solution of NaOH will form sodium carbonate (Na_2CO_3). This would thus enhance the absorption of CO_2 in solution. In literature one finds many systems where aqueous solutions of amines have been used to capture CO_2 (Chapter 5).

In another case: Gas is injected (a suitable scrubber) into a mixture of water + organic substance (non-miscible with water) (alcohol, alkane, etc.). The water/organic liquid system creates a two-phase system. These gas—liquid$_{water}$—liquid$_{immiscible}$ systems have gained interest in the past decade owing to the introduction of homogeneous biphasic catalysis in various reaction systems, for example, hydro-formylation, carbonylation, hydrogenation, and oligomerization (Cents et al., 2001; Cornils, 1999). The main advantage of these systems over catalysis in one phase is the easy separation of the catalyst and the reactants or products.

Gas—liquid$_1$—liquid$_2$ (where: liquid$_1$ is water (solvent) and liquid$_2$ is a solute) systems are further encountered in reaction systems that inherently consist of three phases because there are two (or more) immiscible reactants, reaction products, or catalyst (Falbe, 1980).

Gas—liquid$_1$—liquid$_2$ systems are developed to add increased absorption characteristics and increased specificity. The approach in this kind of process is to add an additional (inert) liquid phase in order to increase the mass transfer rate. This latter approach has also been applied to some biochemical applications (Rols et al., 1990). However, the addition of a second liquid phase can also retard the gas—liquid mass transfer (Yoshida et al., 1970). Some typical and specific CO_2 absorption studies are given in Chapter 5.

Gas absorption in fluids has been investigated for many decades (Yoshida et al., 1970; Cents et al., 2001). The absorption of oxygen was reported in dispersions of water organic fluids (such as kerosene and toluene). There are also reported absorption studies on the effect of a second immiscible liquid on the gas—liquid interfacial area (Mehta & Sharma, 1971) that used a fast reaction, CO_2—NaOH system to study the influence of 2-ethyl-hexanol on the specific gas—liquid interfacial area. It was found that the interfacial area increased owing to a decreased degree of bubble coalescence.

2.3.1 Absorption of CO_2 in Aqueous Solutions

As an absorption/reaction system, CO_2 (gas) absorption in a 0.5 M potassium carbonate/0.5 M potassium bicarbonate buffer solution was used. The carbonate/bicarbonate buffer solution is the continuous liquid in these studies. As the dispersed liquid phase several different organic liquids were added (such as toluene, n-dodecane, n-heptane, and 1-octanol).

The absorption of CO_2 (gas) in a aqueous solution has been analyzed (Cents et al., 2001). The following reactions are taking place with absorption in the bulk liquid phase:

System: CO_2 + POTASSIUM CARBONATE/BICARBONATE BUFFER (0.5 M)

In this system, the following equilibriums are present:

$$CO_2 + H_2O = H_2CO_3 = HCO_3^- + H^+ \tag{2.30}$$

The equilibrium constant, K:

$$K = (k^{-1}) / (k^{+1}) \tag{2.31}$$

with a forward/reverse rate ratio of k^{+1}/k^{-1}. There is another reaction that follows:

$$CO^{-2}_3 + H^+ = HCO_3^-$$ (2.32)

It is known that this reaction can be catalyzed by different additives (such as hypochlorite, arsenide). At high pH = 10, as present in this buffer, there are also present hydroxyl ions (OH^-). CO_2 reacts with OH^-:

$$CO_2 + OH^- = HCO_3^-$$ (2.33)

In this reaction, the equilibrium concentration of CO_2 is dependent on the **carbonate** and **bicarbonate** concentrations (Danckwerts, 1970) (Appendix C).

CHAPTER 3

Surface Chemistry of Solids

3.1 INTRODUCTION

The Earth is surrounded by air (gases: nitrogen, oxygen, CO_2, traces of other gases), and it actively interacts with the surface of Earth (land areas (solid) and oceans/lakes/rivers (water)). This shows that the interfaces of:

gas (air)—solid (Earth)

and

gas (air)—water (oceans, lakes)

and

CO_2 (air)—water (oceans, lakes)

are important phenomena in various everyday systems. In any process where two different phases (in the present case, solid—gas) are involved, the surface chemistry of the substances becomes very important. In everyday life, one experiences a distinct difference between the surface properties of Teflon and other solids. Further, in complex structures such as coal, the surface properties are found to depend on the composition (Keller et al., 1992; Chattoraj & Birdi, 1984; Yang, 2003; Yu et al., 2012; Myers & Monson, 2014; Birdi, 1997, 2003, 2009, 2014, 2016, 2017).

The *surface of a solid* exhibits chemico-physical properties that are very important in many everyday processes. Experiments show that **solid surface tension** is in many aspects similar (qualitatively) to the surface tension of liquids (Appendix A).

Solid surfaces exhibit some specific characteristics that are of much different properties than the liquid surfaces (Adamson & Gast, 1997; Sheludko, 1966; Chattoraj & Birdi, 1984; Birdi, 2004, 2016; Somasundaran, 2006).

In all processes where solids are involved, the primary process is dependent on the surface property of the solid. In some distinct ways the solid surfaces are found to be different than the liquid surfaces (Chattoraj & Birdi, 1984; Birdi, 2016). One finds a large variety of applications where the surface of a solid plays an important role (for example, active charcoal, talc, cement, sand, catalysis, oil and gas reservoirs (shale reservoirs), plastics, wood, glass, clothes and garments, biology (hair, skin, etc.), road surfaces, polished surfaces, friction (drilling technology), earthquakes, earth erosion, etc.). Solids are rigid structures and resist any stress effects. It is thus seen that many such considerations here in the case of solid surfaces will be somewhat different than for liquids. Further, in all porous solids, the flow of gas or liquid oil means that interfaces (liquid—solid) are involved. Hence, in such systems the molecular interactions of surfaces chemistry of the phases are important. Many important technical and natural (such as earthquakes) processes in everyday life are dependent on the rocks (etc.) in the interior of Earth. This thus requires the understanding of the surface forces on solid interfaces:

(solid—gas; solid—liquid; $solid_1$—$solid_2$).

Currently a major example of the surface chemistry of solids has been described based on the classical theories of chemistry and physics (Adamson & Gast, 1997; Birdi, 2002, 2016; Somasundaran, 2006). Another very important example is the corrosion of metals (the interaction of oxygen in air with certain metals: iron/aluminum/copper/zinc/etc.). The process of corrosion initiates at metal surfaces, thus requiring treatments that are based upon surface properties. As described in the case of liquid surfaces, analogous analyses of solid surfaces have been carried out. If one analyzes the surface of a solid at the molecular level, one finds that the molecules at the solid surface are not under the same force field as in the bulk phase (Figure 3.1).

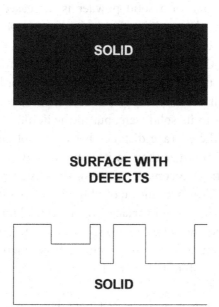

FIGURE 3.1 Solid surface molecular structures: (a) perfect crystal; (b) surface with defects.

Experiments show that the characteristics of solid surfaces are dependent on the nature of the surface structures. In the bulk phase, each solid atom is surrounded symmetrically by near neighboring atoms. However, at the surface, toward the air (for example), there is an asymmetric situation. This asymmetry produces special properties in solids at their surfaces.

Experiments show that the solid surface is the most important characteristic. The differences between perfect surfaces and surfaces with defects are very obvious in many everyday observations. For example, the shine of all solid surfaces increases as the surface becomes smoother. Further, the friction decreases between two solid surfaces as the solid surfaces become smoother. The solids were the first materials that were analyzed at the molecular scale (by using X-ray diffraction, etc.). This led to the understanding of the structures of solid substances and the crystal atomic structure. This is because while molecular structures of solids can be investigated by such methods as X-ray diffraction (SPM), the same analyses for liquids are not that straightforward. These analyses have also shown that surface defects exist at the molecular level.

As pointed out for liquids (Appendix A), one will also consider that when the surface area of a solid powder is increased by grinding (or some other means), then surface energy (energy supplied to the system) is needed. Cement technology is mainly based on this surface character-istic (Birdi, 2010). Of course, because of the energy differences between solid and liquid phases, these processes will be many orders of magnitude different from each other. The liquid state of course retains some struc-ture that is similar to its solid state, but in the liquid state the molecules exchange places. The average distance between molecules in the liquid state is roughly 10% larger than in its solid state (Chapter 1). It is thus desirable at this stage to consider some of the basic properties of liquid solid interfaces. The surface tension of a liquid becomes important when it comes in contact with a solid surface. The interfacial forces that are pres-ent between a liquid and solid can be estimated by studying the shape of a drop of liquid placed on any smooth solid surface (Figure 3.2). The shape (or the contact angle, θ) of the drop of a liquid on different solid surfaces is found to be different.

The balance of forces as indicated have been extensively analyzed, which relates different forces at the solid—liquid boundary and the con-tact angle, θ, as follows (Young's equation 3.1) (Adamson & Gast, 1997; Chattoraj & Birdi, 1984; Birdi, 1997, 2002, 2016):

(Surface tension of solid) =

Surface tension of solid/liquid

+ surface tension of liquid

$$Cos\left(\theta\right) \tag{3.1}$$

$$\gamma_S = \gamma_L Cos\left(\theta\right) + \gamma_{SL} \tag{3.2}$$

Or:

$$\gamma_L Cos(\theta) = \gamma_S - \gamma_{SL} \tag{3.3}$$

where different surface forces (surface tensions) are
γ_S for solid,
γ_L for liquid,
γ_{SL} of solid—liquid interface.

CONTACT ANGLE

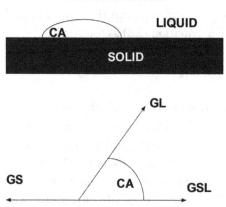

FIGURE 3.2 The state of equilibrium between surface tensions of liquid (GL)—solid (GS)—liquid/solid (GLS)—contact angle (CA).

The relation of Young's equation can be delineated as it follows from simple physical laws. At the equilibrium contact angle, all the relevant surface forces come to a stable state (Figure 3.2).

In this equation only the geometrical force balance is considered, as well as only in the X-Y plane. It is assumed that the liquid does not affect the solid surface structure (in any physical sense). This assumption is safe in most cases. However, only in very special cases, if the solid surface is soft (such as contact lens), will one expect that tangential forces will also need to be included in this equation.

3.2 SOLID SURFACE CHEMISTRY (WETTING PROPERTIES OF SOLID SURFACES)

In all surface phenomena where solid surfaces are involved, the characteristics of the solid surface is of importance. There is no direct procedure to estimate the solid surface property, such as one finds for liquids (i.e., measurement of surface tension).

Another phenomenon of interest is the degree of wetting of a solid surface (Birdi, 2010). The degree of **wetting** when a liquid comes in contact with a solid surface is the most common phenomena in everyday life (washing and detergency, water flow in underground, rain water seepage, cleaning systems, water flow in rocks, water-repellant fabrics, etc.).

The liquid and solid surface interface can be described by considering a classical example. Wetting of solid surfaces is well known when considering the difference between Teflon and metal surfaces. To understand the degree of wetting, between the liquid, L, and the solid, S, it is convenient to rewrite equation (3.3) as follows:

$$Cos (\theta) = (\gamma_S - \gamma_{LS}) / \gamma_L \tag{3.4}$$

which would then allow one to analyze the variation of γ with the change in the other terms. The latter analysis is important because complete wetting occurs when there is no finite contact angle, and thus $\gamma_L <> \gamma_S - \gamma_{LS}$. However, when $\gamma_L > \gamma_S - \gamma_{LS}$, then $Cos (\theta) < 1$, and a finite contact angle is present. The latter is the case when water, for instance, is placed on a hydrophobic solid, such as Teflon, polyethylene or paraffin. The addition of surfactants to water, of course, reduces γ_L; therefore, θ will decrease on the introduction of such surface active substances (Adamson & Gast, 1997; Chattoraj & Birdi, 1984; Birdi, 1997, 2002, 2016). The state of a fluid drop under dynamic conditions, such as evaporation become more complicated (Birdi et al., 1987; Birdi & Vu, 1993). However, in the present it is useful to consider the spreading behavior when a drop of one liquid is placed on the surface of another liquid, especially when the two liquids are immiscible.

The spreading phenomenon was analyzed by introducing a quantity, spreading coefficient, Sa/b, defined as (Harkins, 1952; Adamson & Gast, 1997; Birdi, 1997, 2002, 2016):

$$S_{a/b} = \gamma_a \left(\gamma_b + \gamma_{ab} \right) \tag{3.5}$$

where $S_{a/b}$ is the spreading coefficient for liquid b on liquid a, γ_a and γ_b are the respective surface tensions, and γ_{ab} is the interfacial tension between the two liquids. If the value of $S_{a/b}$ is positive, spreading will take place spontaneously, while if it is negative, liquid b will rest as a lens on liquid a.

However, the value of γ_{ab} needs to be considered as the equilibrium value, and therefore if one considers the system at non-equilibrium, then the spreading coefficients would be different. For instance, the instantaneous spreading of benzene is observed to give a value of $S_{a/b}$ as 8.9 dyn/cm, and therefore benzene spreads on water. However, as the water becomes saturated with time, the value of (water) decreases, and benzene drops tend to form lenses. The short chain hydrocarbons such as hexane and hexene also have positive initial spreading coefficients and spread to give thicker films. Longer chain alkanes, however, do not spread on water

TABLE 3.1 Calculation of Spreading Coefficients, $S_{a/b}$, for Air—Water Interfaces (20°C) (a = air; w = water; o = oil).

Oil	$\gamma_{w/a}$	$\gamma_{o/a}$	$\gamma_{o/w}$	$= S_{a/b}$	Conclusion
$nC_{16}H_{34}$	72.8	30.0	52.1	= 0.3	Will not spread
n-Octane	72.8	21.8	50.8	= +0.2	Will just spread
n-Octanol	72.8	27.5	8.5	= +36.8	Will spread

(OIL = water-insoluble organic substance)

(see Table 3.1), for example, the $S_{a/b}$ for C16 (n-(hexadecane)/water is 1.3 mN/m (dyn/cm) at 25°C.

There are also systems of interest where a solid is placed on water. The spreading characteristics of a solid (polar organic) substance, for example, cetyl alcohol ($C_{18}H_{38}OH$) on the surface of water, has been investigated in some detail (Gaines, 1966; Adamson & Gast, 1997; Birdi, 1997, 2003). Generally, however, the detachment of molecules of the amphiphile into the surface film occurs only at the periphery of the crystal in contact with the air—water surface. In this system, the diffusion of amphiphile through the bulk water phase is expected to be negligible, because the energy barrier now includes not only the formation of a hole in the solid but also the immersion of the hydrocarbon chain in the water. It is also obvious that the diffusion through the bulk liquid is a rather slow process. Furthermore, the value of $S_{a/b}$ would be very sensitive to such impurities as regard spreading of one liquid upon another.

One also finds another example, which is the addition of surfactants (detergents: surface active agents) and how they dramatically affect the liquid's wetting and spreading properties. Thus, many technologies utilize surfactants for control of wetting properties (Birdi, 1997, 2003). The ability of surfactant molecules to control wetting arises from their self-assembly at the liquid—vapor, liquid—liquid, solid—liquid, and solid—air interfaces and the resulting changes in the interfacial energies (Birdi, 1997). These interfacial self-assemblies exhibit structural detail and variation. The molecular structure of the self-assemblies and the effects of these structures on wetting or other phenomena remain topics of extensive scientific and technological interest.

As an example, in the case of oil spills on the seas, these considerations become very important. The treatment of such pollutant systems requires the knowledge of the state of the oil. The thickness of the oil layer will be dependent on the spreading characteristics. The effect on ecology (such as birds, fish, plants) will depend on the spreading characteristics.

Contact angle at liquid₁—sold—liquid₂ interface:
There are also many systems where one finds systems such as

LIQUID$_1$—SOLID—LIQUID$_2$

Or (oil spills on the oceans)

OIL—SOLID—WATER

One typical example is oil spills on the oceans, where one has oil—water—solid. Young's equation at liquid₁—solid—liquid₂ has been investigated for various systems. This is found in such systems where the liquid₁—solid—liquid₂ surface tensions meet at a given contact angle. For example, the contact angle of a water drop on Teflon is 50° in octane (Chattoraj & Birdi, 1984; Adamson & Gast, 1997; Birdi, 2003, 2016) (Figure 3.3):

water   Teflon   octane

In this system the contact angle, θ, is related to the different surface tensions as follows:

$$\gamma_{s\text{-octane}} = \gamma_{water\text{-}s} + \gamma_{octane\text{-}water} \cos\left(\theta\right) \tag{3.6}$$

or

$$\cos\left(\theta\right) = \left(\gamma_{s\text{-octane}} - \gamma_{water\text{-}s}\right) / \gamma_{octane\text{-}water} \tag{3.7}$$

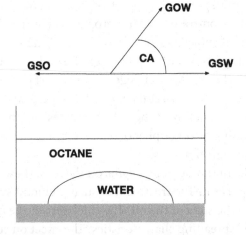

FIGURE 3.3   Contact angle at water—Teflon—octane interface.

This gives the value of $\theta = 50°$, when using the measured values of $\gamma_{s\text{-octane}}$, $\gamma_{water\text{-}s}$ and $\gamma_{octane\text{-}water}$. The experimental value of θ (= 50°) is the same. This analysis showed that the assumptions made in derivation of Young's equation are useful as regards the description of the system.

3.3 SURFACE TENSION OF SOLIDS (SURFACE FORCES)

All natural processes are mainly related to surfaces (e.g., liquid or solid). It is also recognized that the most important property of a surface (solid or liquid) is its capability of interacting with other materials (gases, liquids, or solids). All interactions in nature are governed by different kinds of molecular forces (such as van der Waals, electrostatic, hydrogen bonds, dipole—dipole interactions). Based on various molecular models, the surface tension, γ_{12}, between two phases with γ_1 and γ_2 was given as (Adamson & Gast, 1997; Ross, 1971; Chattoraj & Birdi, 1984; van Oss et al., 1988; Birdi, 1997, 2016):

$$\gamma_{12} = \gamma_1 + \gamma_2 - 2\,\Phi_{12}\,(\gamma_1\,\gamma_2)^2 \tag{3.8}$$

where Φ_{12} is related to the interaction forces across the interface. The latter parameter has been extensively analyzed. It is found that the magnitude of Φ_{12} varies between 0.5 and 1.0 (Birdi, 1997, 2016).

Based on this, one expects that in all liquid—solid interfaces, there will be present different apolar (dispersion) forces + polar (hydrogen-bonding; electrostatic) forces. Hence, all liquids and solids will exhibit γ of different kinds:

Liquid surface tension:

$$\gamma_L = \gamma_{L,D} + \gamma_{L,P} \tag{3.9}$$

Solid surface tension:

$$\gamma_S = \gamma_{S,D} + \gamma_{S,P} \tag{3.10}$$

From this one can conclude that γ_S for Teflon arises only from *dispersion* (γ_{SD}) forces. However, a glass surface shows γ_S, which will be composed of both SD and SP components. Hence, the main difference between Teflon and a glass surface will arise from the SP component of glass. This criterion has been found to be of importance in the case of application of adhesives.

The adhesive used for glass will need to bind to solid with both polar and apolar forces.

The values of γ_{SD}, γ_{SD} and γ_{SD}, for different solids as determined from these analyses are given below.

Solid	γ_S	γ_{SD}	γ_{SP}
Teflon	19	19	0
Polypropylene	28	28	0
Polycarbonate	34	28	6
Nylon 6	41	35	6
Polystyrene	35	34	1
PVC	41	39	2
Kevlar 49	39	25	14
Graphite	44	43	1

As an example: In the case of polystyrene surfaces, it was found that the value of γ_{SP} increased with treatment of sulfuric acid (owing to the formation of sulfonic groups in the surface) (Birdi, 19). This gave rise to increased adhesion of bacteria cells to the surfaces. This criterion is essential for studying bacterial growth in PS dishes.

Most of the information and physical laws of surfaces have been obtained by the studies of liquid—gas or $liquid_1$—$liquid_2$ interfaces. The solid surfaces have been studied in more detail, but the molecular information (at atomic scale) has only taken place during the past decades. The *asymmetrical* forces acting at surfaces of liquids are much shorter than those expected on solid surfaces. This is because of the high energies that stabilize solid structures. Therefore, when one considers solid surface, then the *surface roughness* will need to be considered.

3.3.1 Definition of Solid Surface Tension (γ_{solid})

Based on the observations of contact angles at liquid—solid systems, it thus becomes apparent that these data provide useful information about the interfacial forces (between liquids and solids). As delineated above, the molecules at the surface of a liquid are under tension owing to asymmetrical forces, which gives rise to surface tension. However, in the case of solid surfaces, one may find it difficult to envision this kind of asymmetry as clearly. Although a simple observation might help one to realize that such surface tension analyses are useful.

It is useful to analyze the state of a drop of water (10 μL) as placed on two different smooth solid surfaces, for example, Teflon and glass:

Water—Teflon

Water—Glass

one finds that the contact angles are different (Figure 3.4).

The values of contact angles, θ, are 108° and 35° for Teflon and glass, respectively (Figure 3.4). Since the surface tension of water is the same in the two systems (Figure 3.4), then the difference in contact angles can only arise from the *surface tension of solids* being different.

The surface tension of liquids can be measured directly (Chapter 2; Appendix A). However, this is not possible in the case of solid surfaces. Experiments show that when a liquid drop is placed on a solid surface, the contact angle, θ, indicates that the molecules interact across the interface. This thus means that these data can be used to estimate the surface tension of solids.

3.3.2 Contact Angle (θ) of Liquids on Solid Surfaces

It was explained above, a solid in contact with a liquid leads to interactions related to the surfaces involved (i.e., surface tensions of liquid and

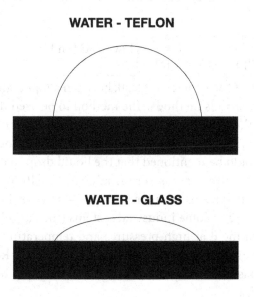

WATER - TEFLON

WATER - GLASS

FIGURE 3.4 Drop of water on the smooth surfaces of Teflon and glass.

solid). The solid surface is being brought in contact with surface forces of the liquid (surface tension of liquid). If a small drop of water is placed on a smooth surface of Teflon or glass, Figure 3.4, one finds that these drops are different.

The reason being that there are three surface forces (tensions) that at equilibrium give rise to a contact angle, θ.

The relationship as given by Young's equation describes the equilibrium between the different surface forces (liquid surface tension; solid surface tension; liquid—solid surface tension) at the three-phase boundary line (and contact angle). It is regarded as if these forces interact along a line. Experimental data show that this is indeed true. The magnitude of θ is thus only dependent on the molecules nearest the interface and is independent of molecules much further away from the contact line.

Further, one defines that:

when θ is less than 90°, the surface is wetting (such as water on glass);

when θ is greater than 90°, then the surface is non-wetting (such as water on Teflon).

The most important one needs to mention is that by treatment of the glass surface with suitable chemicals, the surface can be rendered hydrophobic. This is the same technology as is used in many utensils that are treated with Teflon or similar.

3.4 MEASUREMENT OF CONTACT ANGLES AT LIQUID—SOLID INTERFACE

The magnitude of the contact angle, θ, between a liquid and solid can be determined by various methods. The method to be used depends on the system and on the accuracy required. There are two most common methods: by direct microscope and a goniometer or by photography (digital analyses). It should be mentioned that the liquid drop that one generally uses in such measurements is very small, such as 10 to 100 μL. There are two different systems of interest: liquid—solid or liquid$_1$—solid—liquid$_2$. In the case of some industrial systems (such as oil recovery) one needs to determine θ at high pressures and temperature. In these systems the value of θ can be measured with photography. Recently digital photography has also been used, since these data can be analyzed by computer programs.

TABLE 3.2   Contact Angles, θ, of Water on
Different Solid Surfaces (25°C)

Solid	θ
Teflon (PTE)	108
Paraffin wax	110
Polyethylene (PE)	95
Graphite	86
AgI	70
Polystyrene (PS)	65
Glass	30
Mica	10

It is useful to consider some general conclusions from these data. One defines a solid surface as *wetting* if the θ is less than 90°. However, a solid surface is designated as *non-wetting* if θ is greater than 90°. This is a practical and semi-quantitative procedure. It is also seen that water, owing to its hydrogen bonding properties, exhibits a large θ on non-polar surfaces (PE, PTE, PE). However, one finds lower θ values on polar surfaces (glass, mica).

Furthermore, one may consider how **charged solid surfaces** will exhibit properties that will be related to contact angles. Metal surfaces will exhibit varying degrees of charges at the surfaces. Biological cells will exhibit charges that will affect adhesion properties to solid surfaces. However, in some applications one may change the surface properties of a solid with chemical modifications of the surface. For instance, polystyrene (PS has some weak polar groups at the surface). If one treats the surface with H_2SO_4, which forms sulfonic groups, this leads to values of θ lower than 30° (which is found to be dependent on the time of contact between sulfuric acid and the PS surface). This treatment (or similar) has been used in many applications where the solid surface is modified to achieve a specific property. Since only the surface layer (a few molecules deep) is modified, the solid properties bulk do not change. These analyses show the significant role of studying the contact angle of surfaces in relation to the application characteristics. A more rigorous analysis of these systems has been given in the literature. Another very important study is the characteristics of coal surface and related applications (Birdi, 2016).

The magnitude of the contact angle of water (for example) is found to vary depending on the nature of the solid surface. The magnitude of θ is almost 100° on a waxed surface of car paint. The industry strives to create such surfaces to give $\theta > 150°$, the so-called super-hydrophobic surfaces. The large θ means that water drops do not wet the car polish and are easily blown off by wind. The car polish also is designed to leave a highly *smooth* surface.

In many industrial applications one is both concerned with smooth and rough surfaces. The analyses of θ on *rough surfaces* will be somewhat complicated than on smooth surfaces. The liquid drop on a rough surface, Figure 3.5, may show the *real* θ (solid line) or some lower value (*apparent*) depending on the orientation of the drop. Surface thermodynamics analyses show that there is only the *real* θ that is of interest.

However, no matter how rough the surface, the surface (molecular) forces will be the same as those that exist between a solid and liquid. In other words, at micro-scale, surface roughness has no effect on the balance of forces at the liquid—solid and contact angle. The surface roughness may show contact angle *hysteresis* if one makes the drop move, but this will arise from other parameters (e.g., wetting and de-wetting) (Birdi, 2016). Further, in practice the surface roughness is not easily defined. A *fractal* approach has been used to achieve a better

understanding (Feder, 1988; Birdi, 1993). In various industries one finds many systems where contact angle analyses are found to be useful (oil/gas reservoirs, oil spills on oceans, washing and cleaning, erosion by rain drops, etc.). Table 3.3 gives some unusual data for different systems.

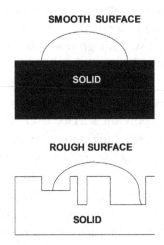

FIGURE 3.5   Analysis of contact angle of a liquid drop on a rough solid surface.

TABLE 3.3   Contact Angle Data of Different Liquid—Solid Systems

System		Contact Angle
Liquid	γ	
Liquid Na	222	66°
(100°C)—Glass		
Hg (100°C)—Glass	460	143°

It can be safe to conclude that even though Young's equation was based on a simple model, it has been found to give useful analyses for a variety of systems. A typical data of cos (θ) versus various liquids on Teflon gave an almost straight line data plot (Birdi, 1997, 2003). These data can be analyzed by the following relation:

$$\cos (\theta) = k_1 - k_2 \, \gamma_L \tag{3.11}$$

This can also be rewritten as thus:

$$\text{Cos} (\theta) = 1 - k_3 \, (\gamma_L - \gamma_{cr}) \tag{3.12}$$

where γ_{cr} is the critical value of γ_L at $\text{Cos}(\theta)$ equal to 0. The values of γ_{cr} have been reported for different solids using this procedure (Adamson & Gast, 1997; Birdi, 1997, 2016).

The magnitude of γ_{cr} for Teflon of 18 mN/m thus suggests that -CF$_2$- groups exhibit this low surface tension. The value of γ_{cr} for -CH$_2$-CH$_3$- alkyl chains gave a higher value of 22 mN/m than for Teflon. This is also as one would expect. Indeed, from experience one also finds that Teflon is a better water-repellent surface than any other material. Hence, one can conclude that contact angle analyses can provide much useful information as regards the solid surface tension.

3.5 ADSORPTION OF GAS ON SOLID SURFACE (SPECIFIC ROLE OF SOLID SURFACE TENSION)

As already mentioned above, the most important solid surface property is its interaction with gases or liquids. The adsorption of gas on a solid would thus be expected to be related to the surface tension. The molecules in gas are moving very fast, but on adsorption (gas molecules are more or less fixed) there will be thus a large decrease in kinetic energy (thus decrease in entropy, ΔS). The surface properties of coal have been extensively studied with respect to its adsorption characteristics (Birdi, 2016). The surface tension of coal was found to be related to its adsorption properties. Another very important study of shale rock has been found to be useful with regard to the fracking process (Birdi, 2017).

Molecules in the gas phase move much larger distances than when adsorbed on a solid surface. Adsorption takes place spontaneously, which means

$$\Delta G_{ad} = \Delta H_{ad} - T\,\Delta S_{ad} \tag{3.13}$$

If ΔG_{ad} is *negative,* this suggests that ΔH_{ad} is negative (exothermic). The adsorption of gas can be of different types. The gas molecule may adsorb as a kind of *condensation* process, it may under other circumstances react with the solid surface (chemical adsorption or chemisorption). In the case of chemo-adsorption, one almost expects a chemical bond formation. On carbon, while oxygen adsorbs (or chemisorb), one can desorb CO or CO_2. The experimental data can provide information on the type of adsorption. On porous solid surfaces, the adsorption may give rise to *capillary condensation*. This indicates that porous solid surfaces will exhibit some specific properties. The most common adsorption process in industry one finds is the case of catalytic reactions (for example, formation of NH_3 from N_2 and H_2). It is thus apparent that in gas recovery from shale the desorption of gas (mainly methane, CH_4) will be determined by surface forces (Birdi, 2017).

The surface of a solid may differ in many ways from its bulk composition. Especially, solids such as commercial carbon black may contain minor amounts of impurities (such as aromatics, phenol, carboxylic acid). This would render surface adsorption characteristics different than on pure carbon. It is therefore essential that in industrial production one maintains quality control of surface from different production batches. Otherwise, the surface properties will affect the application.

The silica surface has been considered to exist as O-Si-O as well as hydroxyl groups formed with water molecules. The orientation of the different groups may also be different at surface.

Carbon black has been reported to possess different kinds of chemical groups on its surface. These different groups are aromatics, phenol, carboxylic, etc. These different sites can be estimated by comparing the adsorption characteristics of different adsorbents (such as hexane and toluene). When any *clean* solid surface is exposed to a gas, the latter may adsorb on the solid surface to a varying degree (as found from experiments). It has been recognized for many decades that gas adsorption on solid surfaces does not stop at a monolayer state. Of course, more than one layer (multilayer) of adsorption will take place only if the pressure is reasonably high. Experimental data show this when the volume of gas adsorbed, v_{gas}, is plotted against P_{gas} (Figure 2.7).

From experimental analyses it has been found that there exist six typical different kinds of adsorption states (Figure 2.6). These adsorption isotherms were classified based on extensive measurements and analyses of v_{gas} versus p_{gas} data (Chapter 2). For example:

Type I: These are obtained for Langmuir adsorption. Adsorption of ammonia (NH_3) on charcoal is a system where this type of isotherm is observed.

Type II: This is the most common type where multilayer surface adsorption is observed.
Adsorption of nitrogen (N_2) on silica is this type of isotherm.

Type III: This is a somewhat special type with almost only multilayer formation, such as nitrogen adsorption on ice.

Type IV: If the solid surface is porous, then this is similar to type II. The adsorption data of benzene ($C_6 H_6$) on silica were found to fit this isotherm.

Type V: On porous solid surfaces, type III. Water adsorption on charcoal data was found to be this type of isotherm.

The pores in a porous solid surface are found to vary from 2 nm to 50 nm (*micropores*).

Macropores are designated for larger than 50 nm. *Mesopores* are used for the 2 to 50 nm range.

3.5.1 Gas Adsorption Measurement Methods

The gas adsorption can be measured by using various techniques, depending on the gas—solid system (and the experimental conditions: temperature, pressure, etc.).

Volumetric Change Methods

The change in volume of gas during adsorption is measured directly in principle, and the apparatus is comparatively simple (Figure 3.6).

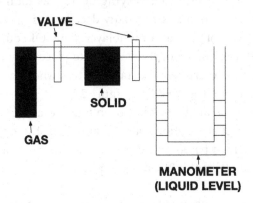

FIGURE 3.6 A typical gas adsorption on solid surface apparatus.

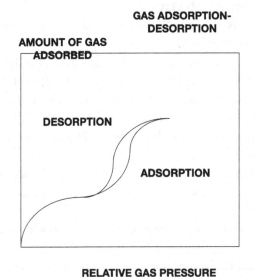

FIGURE 3.7 Adsorption/desorption isotherm of gas—solid system.

One can use a mercury (or any other suitable liquid) reservoir in the manometer, and the burette is used to control the levels of mercury in the apparatus. Calibration involves measuring the volumes of the gas (v_g) tubings and of the void space (Figure 3.7). All pressure measurements are made with the right arm of the manometer set at a fixed zero point so that the volume of the gas lines does not change when the pressure changes. The apparatus, including the sample, is evacuated, and the sample is heated in order to remove any previously adsorbed gas. Generally, an inert gas such as helium is usually used for the calibration, since it exhibits very low adsorption on the solid surface. After helium is pushed into the apparatus, a change in volume is used to calibrate the apparatus, and the corresponding change in pressure is measured. A different gas (such as nitrogen) is normally used as the adsorbate if one needs to estimate the surface area of a solid. The gas is cooled by liquid nitrogen. The tap to the sample bulb is opened, and the drop-in pressure is determined. In the surface area calculations, one uses a value of 0.162 nm^2 for the area of an adsorbed nitrogen molecule.

From simple calculations, these data are used to estimate the quantity surface area per gram of solid.

Gravimetric Gas Adsorption Methods
It is obvious that in general, the amount of gas adsorbed on any solid surface will be of a very small magnitude. An appropriately sensitive microbalance is generally used to measure the adsorption isotherm. The sensitivity is very high since only the difference in weight change is measured. These microbalances can measure weight differences in the range of nanograms (10^{-9} gm) to milligrams (10^{-3} gm). With such extreme sensitivity, it is possible to measure the weight change caused by the adsorption of a single monolayer on a solid if the surface area is large. The normal procedure is to expose the sample to the adsorbate gas at a certain pressure, allowing sufficient time for equilibrium to be reached and then determining the mass change. This is repeated for a number of different pressures, and the number of moles adsorbed as a function of pressure plotted to give an adsorption isotherm.

Microbalances (stainless steel) can be made to handle pressures as high as 120 Mpa (120 atm), since gases that adsorb weakly or boil at very low pressures can still be used.

Gas Adsorption on Solid Surfaces (Langmuir Theory)
The simplest theory was based on the assumption that only *one layer* of gas molecule adsorbs. A monolayer of gas adsorbs in the case of where there are only a given number of adsorption sites for only a monolayer. This is the simplest adsorption model. The amount adsorbed, N_s, is related to the monolayer coverage, N_{sm}, as follows (Chapter 2):

$$N_s/N_{sm} = \mathbf{a}p/(1+\mathbf{a}p) \tag{3.14}$$

where p is the pressure and $\mathbf{a}$ is dependent on the energy of adsorption. This equation can be rearranged:

$$p/N_s = \left(1/(\mathbf{a}N_{sm}) + p/N_{sm}\right) \tag{3.15}$$

From the experimental data, one can plot p/N_s versus p. The plot will be linear, and the slope is equal to $1/N_{sm}$. The intersection gives $\mathbf{a}$. Charcoal is found to adsorb 15 mg of N_2 as a mono-layer. Another example is that of adsorption of N_2 on a mica surface (at 90 K). The following data were found:

Pressure/Pa volume of gas adsorbed (at STP)

0.3	12
0.5	17
1.0	24

(standard temperature and pressure, STP)

In this adsorption model one assumes (Chapter 2)

that the molecules adsorb on definite sites;

that the adsorbed molecules are stable after adsorption.

The surface area of the solid can be estimated from the plot of p/N_s versus p. Most data fit this equation under normal conditions and are therefore widely applied to analyze adsorption process.

Langmuir adsorption is found for the data of nitrogen on mica (at 90 K). The data were found to be

$p = 1/Pa \qquad 2/Pa$

$Vs = 24 \text{ mm}^3 \qquad 28 \text{ mm}^3$

This shows that the amount of gas adsorbed increases by a factor of $28/24 = 1.2$ when the gas pressure increases two-fold.

Various Gas—Solid Adsorption (Desorption) Equations

Other isotherm equations begin as an alternative approach to the developed equation of state for a two-dimensional ideal gas (Chapter 2). The assumption one has made in deriving the equation of state form of the isotherm equation is based on gas theory. As mentioned earlier, the ideal equation of state is found to be as follows (Appendix A):

$$\Pi A = k_B T \tag{3.16}$$

Where Π is the surface pressure of the adsorbed gas, A is the area/molecule of the gas, and k_B is the Boltzmann constant. In combination with the Langmuir equation one can derive the following relation between N_s and p:

$$N_s = K p. \tag{3.17}$$

where **K** is a constant. This is the well-known Henry's law relation, and it is found to be valid for most isotherms at low relative pressures. In those situations where the ideal equation 3.16 does not fit the data, the van der Waals equation type of corrections have been suggested.

The *adsorption—desorption process* is of interest in many systems (such as cement, shale reservoir, CCS). The water vapor may condense in the pores of cement, after adsorption under certain conditions. This may be studied by analyzing the adsorption—desorption data (Figure 35.9). This criterion is of importance where captured (adsorbed) gas is needed to be desorbed (extracted), as in the case of CCS technology (Chapter 4 and 5). Therefore, in CCS one would use solids that exhibit better desorption characteristics (Chapter 5).

Multilayer gas adsorption: In another example, the shale fracking process for gas recovery from shale deposits is a process where adsorbed gas is released. In some systems adsorption of gas molecules proceeds to higher levels, where multilayers are observed. From data analyses one finds that multi-layer adsorption takes place.

The BET equation has been derived for multilayer adsorption data. The enthalpy involved in multilayers is related to the differences between the adsorption energies and was defined by **BET** theory as thus:

$$E_{BET} = \exp[(E_1 - E_v)/RT] \tag{3.18}$$

where E_1 and E_v are enthalpies of adsorption and desorption. The **BET** equation thus after modification of the Langmuir equation becomes thus:

$$p/(N_s(p - p)) = 1/E_{BET} N_{sm} + [(E_{BET} - 1)/(E_{BET} N_{sm})] (p / po) \tag{3.19}$$

From a plot of adsorption data, the left hand side of this equation versus relative pressure (p/po) allows one to estimate N_{sm} and E_{BET}. The magnitude of E_{BET} is found to give either data plots that are type III or II. If the value of E_{BET} is low, which means the interaction between adsorbate and solid is weak, then type III plots are observed. This has been explained as arising from the fact that if $E_1 > E_v$, then molecules will tend to form multilayers in patches rather than adsorb on the naked surface. If there exists strong interaction between the gas molecule and the solid, $E_1 < E_v$, then type II plots are observed. The mono-layer coverage is clearly observed at low values of p/p_o.

CHAPTER 4

Carbon Capture and Sequestration (CCS) Technology (Basic Remarks)

4.1 INTRODUCTION

The evolution of the solar system has been going on for over a period of few billions of years and has brought the state of the Earth in pseudo equilibrium with its surroundings. That means the sun—atmosphere—Earth system has reached a state of pseudo equilibrium (as regards chemical and bio-chemical equilibrium, temperature and pressure equilibrium, gravity forces, rain fall and rivers, earthquakes, etc.). For example, the average global temperature of the Earth is in equilibrium with the effective heat reaching from the sun. However, interaction between the Sun and the Earth has to pass through the atmosphere (air, moisture (water molecule), clouds, dust particles, etc.). The pressure and temperature of air (atmosphere) is different at different heights from the surface of the Earth (Chapter 1). Furthermore, each component of gas in air has a specific role in the evolution process on Earth. For example, oxygen and CO_2 are very important molecules for the life cycle. Nitrogen is an inert gas and has very little direct interaction with everyday life. As regards CO_2, it has some very significant interactions with life on Earth. It is also a toxic gas, and around a CO_2 concentration of 14%, no life can exist. During the evolutionary process, the concentration has remained below a few thousand ppm (<1%). The eco-system in the oceans consists of various entities, biological and non-biological. The non-biological evolution in the oceans is controlled primarily by water. Evolution based on the chemical properties of water is very significant as regards the life on Earth. The solubility of CO_2 in water is one of the most important systems, which is evolved into a complex equilibrium. CO_2 as found in air is in equilibrium with CO_2 in water. However, CO_2 in water is found to react with water and other metal ions (such as calcium, magnesium, etc.), thus creating a very complex carbonate (i.e., HCO_3/CO_3) eco-system in the oceans (Figure 4.1).

Over billions of years, this carbonate system has reached an equilibrium with its surroundings (i.e., air (that is, CO_2 in air)—water—carbonate salts in the oceans) (Rackley, 2010; David et al., 2008). This means, as also is evident for other eco-systems on Earth, the latter equilibrium is very large and does not fluctuate appreciably within short time periods. This observation is also evident from the CO_2 data over many thousands of years. These data indeed show a very stable content with slow fluctuations. This means that the atmosphere (especially its composition) is a complex system. The composition of air near the

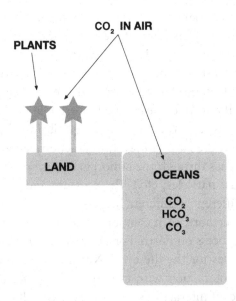

FIGURE 4.1 Carbon dioxide (CO_2) equilibria: air—land areas (plants)—oceans/lakes (carbonate).

surface of the Earth has been monitored for many decades. The concentration of nitrogen (78%) and oxygen (21%) is found to have remained fairly constant over many decades. However, one has analyzed and monitored the CO_2 concentration (varying from 280 ppm to the current 400 ppm) variation over a century. This has been carried out since it is known to be a GHG.

There are extensive analyses in the literature as regards the change in composition of GHGs in the atmosphere. For instance, the current concentration of CO_2 is ca. 400 ppm (0.04%). However, analyses indicate that by 2100 this value may increase to 1000 ppm (IPCC, 2007; Dubey et al., 2002). This is argued from the fact that current fossil fuel burning adds (ca. 5 ppm/year) CO_2 to the atmosphere. Therefore, extensive studies are found in the literature, which are related to capture and control of CO_2 from air or during combustion of fossil fuels at plants (flue gases), in order to mitigate the CO_2 concentration (increase) phenomenon. It is important to mention that there are studies that show that the concentration of CO_2 has some significant characteristics:

150 ppm: plants may stop growing;

280 ppm: pre-industrial average (minimum CO_2 concentration needed for plant growth and to sustain life on Earth);

410 ppm: current value;

1000–2000 ppm: variation some thousands year ago.

Carbon capture and storage (**CCS**) (or carbon capture and sequestration or carbon control and sequestration) (Albo et al., 2010; Oh, 2010; Bishnoi & Rochelle, 2000; Yu et al., 2012; Fanchi & Fanchi, 2016; IPCC, 2011, 2014a) is described as the process of capturing waste CO_2 from large point sources (such as fossil fuel power plants), transporting it to a storage site, and depositing it where it will not enter the atmosphere, normally an underground geological formation. The aim is to prevent the release of large quantities of CO_2 (anthropogenic) into the atmosphere (from fossil fuel use in power generation and other industries). It is a potential means of mitigating the contribution of fossil fuel emissions to global warming. Generally, power plants deliver 500–1000 MW of electric power. It is estimated that1000 MW plants produce the following amounts of CO_2: coal, ca. 7 Mt/year of CO_2; oil, 5 Mt/year of CO_2; gas, 3 Mt/year of CO_2.

Furthermore, it is known that CO_2 is soluble in water. This means that CO_2 is soluble in the oceans (lakes, rivers). The Earth is covered by more than 70% of the Earth's surface. Oceans thus are found to contain a very large amount of CO_2. Furthermore, CO_2 dissolves in water and forms bi-carbonate (HCO_3^-) and carbonate (CO_3^-) ions. Since oceans (with depths of several kilometers) and lakes comprise vast quantities of water, it is obvious that there will be present large amounts of CO_2/bi-carbonate/carbonate. CO_2 as present in oceans (lakes), however, does not contribute to the GHG phenomenon.

CO_2 solubility equilibria in oceans/lakes (water): The CO_2 solubility equilibria in oceans/lakes is very extensively studied in literature (Schulz et al., 2006; Rackley, 2010). A brief mention is made here, in order to explain the CO_2 concentration in atmosphere during evolution.

The following equilibria are present: $CO_{2gas} — CO_{2aqueous}$:

CO_2 (gas: in air)// in equilibrium//CO_2 (gas: in water)

$$CO_{2gas} = CO_{2aqueous} \tag{4.1}$$

The equilibrium constant is

$$K_{CO2} = \left(CO_{2aqueous}\right) / \left(x_{CO2}\right) CO_{2gas} \left(p_{CO2}\right) \tag{4.2}$$

where K_{CO2} is the equilibrium constant; $CO_{2aqueous}$ is the concentration of CO_2 in water (mol/kg); x_{CO2} is the mole fraction of CO_2 (in air); CO_{2gas} is the concentration in air; p_{CO2} is the partial pressure.

$CO_{2aqueous}$ and water interact (instantly):

$$CO_{2aqueous} + H_2O = H^+ + HCO_3^- \tag{4.3}$$

With an equilibrium constant thus:

$$K_{HCO3} = ((H+) (HCO_3^-) / (CO_{2aqueous})) \tag{4.4}$$

And dissociation equilibrium constant thus:

$$K_{CO3} = \left(H^+\right)\left(CO_3^-\right) / \left(HCO_3^-\right) \tag{4.5}$$

where K_{CHCO3} is the equilibrium constant, (H^+) is the concentration of hydrogen ion, x_{CO2} is the mole fraction, CO_{2gas} is the concentration in air, p_{CO2} is the pressure.

These equilibria show that during evolution the concentrations of CO_2 in air and oceans have reached an equilibrium. Anthropogenic CO_2 (i.e., fossil fuel combustion) is very small as compared to the carbon stored in the oceans/lakes. However, the carbonate/bicarbonate equilibrium is capturing parts of anthropogenic CO_2 from air. The total carbon content in oceans is thus the sum of all the different quantities:

$$\left[CO_{2aqueous} + CO_3^{-2} + HCO_3^- \right] \tag{4.6}$$

It is also found from these equilibria that any change in the concentration of CO_2 in air would result in a change in the components of these equilibria. The carbonates in the oceans thus stabilize the average concentration of CO_2 in air. The photosynthesis process (which captures CO_2) is much slower than the equilibria in oceans. However, the most significant observations from these data are as follows:

(1) The CO_2 concentration in air is controlled mainly by the equilibria in oceans. This arises from the fact that the total inorganic carbon in the oceans (e.g., $CO_{2aqueous}$ + HCO_3 + CO_3, with ratios of approximately 1:10:100) is a few decades times larger than the CO_2 in air.

(2) The CO_2 concentration in air is thus always comparatively **low** because of these equilibria. This may be due to some evolutionary processes which have brought about these constraints on the concentration of CO_2. Therefore, the data also indicate that the concentrations of CO_2 have always been fluctuating around 1000 ppm (0.1%) over the past many thousands of years.

(3) Photosynthesis and forests (trees/plants, etc.) capture CO_2 from air, and the equilibria under (1) keep the minimum CO_2 in the air.

These data are merely mentioned here, since some studies have investigated the possibility of carbon storage in deep oceans (which is stratified). Analyses of CO_2 in oceans show that because of a lack of mixing (at very deep structures), this is a pseudo equilibrium (Schulz et al., 2006).

After the CO_2 is recovered from flue gas (or directly from air), it needs to be stored or used in some application (such as food/drinks/enhanced oil recovery (EOR), etc.). As regards CO_2 storage, it has been injected into geological formations for several decades for various purposes, including enhanced oil recovery. CO_2 has also been stored in suitable old reservoirs.

Storage of the captured CO_2 is currently also being envisaged either in deep geological formations, or in the form of mineral carbonates (as M_xCO_3). It is also possible technically to store CO_2 in deep oceans. However, the CO_2 solubility will lead to acidification. It is also known that owing to the large deep structures of oceans, there is not complete mixing of the CO_2 (as absorbed from air). Intact, one finds large stratifications and the oceans are only at pseudo-equilibria with CO_2 in air.

Furthermore, CCS applied to a modern conventional power plant could reduce CO_2 emissions to the atmosphere by approximately 80–90% compared to a plant without CCS. It is (IPCC, 2005a, b) estimated that the economic potential of CCS could be between 10% and 55% of the total carbon mitigation application until year 2100.

CO_2 can be captured out of air or fossil fuel power plant flue gas by adsorption (or carbon scrubbing), membrane gas separation, or other adsorption technologies (such as cryogenic process, mineral capture, solution storage in deep oceans, etc.). Absorption of CO_2 by amines is currently the leading carbon-scrubbing technology. Capturing and compressing CO_2 may increase the energy needs of a coal-fired CCS plant by 25–40%. Carbon capture is obviously a costly addition to the industrial sector where fossil fuels are burned (combustion process). These and other system costs are estimated to increase the cost per watt energy produced by 21–91% for fossil fuel power plants.

It has been suggested that, with the development of relevant research, development, and deployment, sequestered coal-based electricity generation in 2025 may cost less than un-sequestered coal-based electricity generation today. Geological formations are currently considered the most promising sequestration sites. The National Energy Technology Laboratory (NETL, 2007, 2015) reported that North America has enough storage capacity for more than 900 years' worth of CO_2 at current production rates (NETL, 2007). A general problem is that long term predictions about submarine or underground storage security are very difficult and uncertain, and there is still the risk that CO_2 might leak into the atmosphere (Phelps et al., 2015).

However, the technology of capturing CO_2 is reported to be most effective at point sources, such as large fossil fuel or biomass energy facilities, industries with major CO_2 emissions, natural gas processing, synthetic fuel plants and fossil fuel-based hydrogen production plants. Extracting CO_2 from air is also viable process.

Flue gas from the combustion of coal in oxygen has a relatively large concentration of CO_2, about 10–15% CO_2, whereas natural gas power plant flue gas is about 5–10% CO_2 (McDonald et al., 2015). Therefore, it is more energy and cost efficient to capture CO_2 from coal-fired power plants. Impurities in CO_2 streams, like sulfur and water, could have a significant effect on their phase behavior. In those systems where CO_2 impurities are present, one will need a scrubbing separation process to remove these from the flue gas.

FOSSIL FUEL COMBUSTION PROCESS

Since combustion of fossil fuels produces CO_2, it is obvious that this process has been investigated in recent reports. For example, it has been suggested that, by gasifying coal, it is possible to capture approximately 65% of CO_2 embedded in it and sequester it in a solid form. The **combustion** process of fossil fuels is given as thus:

$$\text{Fossil fuel} + \text{oxygen (from air)} = CO_2 + H_2O + \text{diverse pollutant gases} \tag{4.7}$$

The diverse pollutants gases that are produced are NO_x, CO, and SO_2. In general, one finds three different configurations of technologies for carbon capture:

post-combustion, pre-combustion, and oxy-fuel combustion.

In post combustion capture, the CO_2 is removed after combustion of the fossil fuel —this is the scheme that would be applied to fossil-fuel burning power plants. Here, CO_2 is captured from flue gases at power stations or other large point sources. The technology is currently at an advanced stage and is used in other industrial applications, although not at the same scale as might be required in a commercial-scale power station. Post-combustion capture is most popular in research because existing fossil fuel power plants can be retrofitted to include CCS technology in this configuration (Sumida et al., 2012).

The technology for pre-combustion is widely applied in fertilizer, chemical, gaseous fuel (H_2, CH_4), and power production. In these cases, the fossil fuel is partially oxidized, for instance in a gasifier. The resulting syngas (CO and H_2) is shifted into CO_2 and hydrogen (H_2). The resulting CO_2 can be captured from a relatively pure exhaust stream. Hence, hydrogen

(H_2) can now be used as fuel; the CO_2 is removed before combustion takes place. There are several advantages and disadvantages when compared to conventional post-combustion CO_2 capture (Masel, 1996) and integrated gasification combined cycle for carbon capture storage.

CO_2 is captured and removed after combustion of fossil fuels. This scheme is applied to new fossil fuel-burning power plants or to existing plants where re-powering is an option. The capture before expansion, that is, from pressurized gas, is a normal process in almost all industrial CO_2 capture processes, at the same scale as will be required for utility power plants.

4.2 CCS TECHNOLOGY AND DIFFERENT METHODS OF CAPTURE

Besides the main methods of carbon capture, for example, adsorption on solids and absorption in fluids (Chapter 2 and 5), there are some other procedures that one finds in the current literature. These will be delineated here briefly.

Direct Capture of CO_2 from Air

Direct air capture from air is a process of removing CO_2 directly from the ambient air (as opposed to from point sources). Combining direct air capture with carbon storage could be useful as a CO_2 removal technology and as such would constitute a form of climate engineering if deployed at large scale.

Although it is out of the scope of this book, only a short mention is given as regards the cost of such a CCS process, which is estimated to be about two thirds of the total cost of CCS and therefore limits the wide-scale deployment of CCS technologies.

Another method that is reported is called chemical looping combustion (CLC). Chemical looping uses a metal oxide as a solid oxygen carrier. Metal oxide particles react with a solid, liquid, or gaseous fuel in a fluidized bed.

$$\text{AIR }(CO_2\,0.04\%)\text{-------CAPTURE TECHNOLOGY-----PURE }CO_2 \quad (4.8)$$

Accordingly, some technological methods have been investigated for such direct air capture.

Furthermore, an economic estimate of this carbon recovery application (in 2018) estimated the cost at ca. 150 USD per ton (1000 kg) of atmospheric CO_2 captured. It is expected that such estimates will decrease as better techniques are developed.

In these studies, the following are the most significant methods:

(1) using alkali and alkali-Earth hydroxides;
(2) carbonation;
(3) organic—inorganic hybrid sorbents consisting of amines supported in porous adsorbents.

The concentration of CO_2 in the atmosphere is highly diluted compared to point-source CO_2 capture. Accordingly, the capture costs are expected to be high. However, the carbon recovery costs will decrease as the importance of its effect on climate change increases.

In the current literature one finds different forms have been conceived for permanent storage of CO_2. These forms include gaseous storage in various deep geological formations (including saline formations and exhausted gas fields), and solid storage by reaction of CO_2 with metal oxides to produce stable carbonates.

Geological storage of CO_2: This technology is about handling of captured CO_2. The captured CO_2 has been stored under varying procedures. This procedure has been also called *geo-sequestration*; this method involves injecting CO_2, generally in supercritical form, directly into underground geological formations. Oil fields, gas fields, saline formations, unmineable coal seams, and saline-filled basalt formations have been suggested as storage sites (Birdi, 2017).

CAPTURED CO_2—STORAGE (OIL FIELD/SALINE FORMATIONS/ UNMINEABLE COAL SEAMS)

Various physical (e.g., highly impermeable caprock) and geochemical trapping mechanisms would prevent the CO_2 from escaping to the surface.

CO_2 has been used in oil recovery processes (EOR). This process is sometimes used where CO_2 is injected into declining oil fields to increase oil recovery. Approximately 200 million tonnes of CO_2 are injected annually in the declining oil fields. This process is considered to be attractive because the geology of hydrocarbon reservoirs is generally well understood, and storage costs may be partly offset by the increased oil recovery. Oil recovery from reservoirs is a multistep process. Most of the recovery is primarily based on original pressure in the reservoir (producing around

20% of oil in place). Enhanced oil recovery (EOR) is a process that is used to increase the amount of crude oil that can be extracted from an oil field (Birdi, 2017). In carbon capture and sequestration enhanced oil recovery (CCS EOR), CO_2 is injected into an oil field to recover oil that is often never recovered using more traditional methods.

CO_2 INJECTION IN OIL RESERVOIRS—ENHANCED OIL RECOVERY (EOR)

Most crude oil production technology in reservoirs consist of three distinct phases: primary, secondary, and tertiary (or enhanced) recovery. During primary recovery only about 10% of a reservoir's original oil in place is typically produced. Secondary recovery techniques extend a field's productive life generally by injecting water or gas to displace oil and drive it to a production wellbore, resulting in the recovery of 20% to 40% of the original oil in place.

4.2.1 Carbon Capture Methods

There are various approaches to determine the feasible application of technologies that can be viable for capturing and storage in the current literature (Herzog, 2016). In the following some of these reports are given.

CARBON DIOXIDE GAS-HYDRATE CAPTURE

The structures of water and ice are found to be abnormal, as compared to other liquid—solid systems. For instance, ice is about 10% lighter than water (that is why icebergs float on water). The simple reason for this characteristic is that the molecular structure of ice is such that the volume per water molecule is larger than in water (at 0°C). This open structure of ice is found to capture some gas molecules. These ice—gas structures are called gas-hydrates (Tanford, 1980; Birdi, 2017). In nature, one finds that under suitable pressure and temperature, gas hydrates are present (in very large quantities). Most of these are methane hydrates. They are found along continental margins worldwide and are located at and above the gas hydrate stability zone. It is thus likely that few, if any, of these methane discharge events along continental slopes are perhaps related to anthropogenic climate change and global warming.

Carbon capture using algae growth: It is found that a large amount of industry using micro-algae produce viable foods (Herzog, 2016). This is found to amount to over 5000 tons of food per year. It is also possible to convert the algae to bio-fuels.

Carbon capture using ocean fertilization: It has been suggested that by enhancing the fertilization in oceans (by using iron), the growth of marine phytoplankton. This will be expected to partially sink to the bottom of oceans/lakes and will remain for many decades.

Carbon capture by mineral (e.g., carbonates) storage: It is already known that many minerals as found in oceans are made up of carbonate (M_xCO_3), which uses (captures) CO_2. An example is calcium magnesium silicates. The reaction between the magnesium silicate (serpentine) is as follows:

$$Mg_3SiO_2(OH)_4 + 3\ CO_{2gas} = 3\ MgCO_3 + 2\ SiO_2 + 2\ H_2O \tag{4.9}$$

CO_2 sources and sinks: CO_2 concentration in air is currently 400 ppm. It is appropriate to consider briefly the different CO_2 sources and sinks parameters. The combustion of fossil fuels worldwide is reported to produce approximately 30 Gt of CO_2 per year. Deforestation of tropical regions accounts for an additional 4 Gt CO_2 per year. For the CO_2 natural (carbon) content cycle and also the related terrestrial and ocean water CO_2 sinks (due to solubility in water), the annual increase in CO_2 emissions are reported to be approximately 15 Gt CO_2 per year. This quantity is almost equivalent to 2 ppm per year (concentrations of CO_2 in air). Fossil fuel-based emissions of CO_2 may be sourced from both stationary (e.g., power plant) and non-stationary systems (e.g., automobile (transport), etc.). The amount emitted is approximately 13 Gt CO_2 per year on average from large stationary sources globally. In addition to CO_2 emissions generated from the oxidation of fossil fuels, flue gases may also be sourced as a result of a chemical process. Although these emission sources represent a minor portion of total anthropogenic emissions, the chemical processing method currently used may be required for the formation of a useful product, such as cement or steel. Therefore, because of the difficulty of replacing CO_2-generating chemical processes with others that are absent of CO_2, it is crucial that these emission sources are not disregarded. The majority of fossil fuel oxidation (combustion) processes produce CO_2 emissions, whereas a fraction of emissions are generated from chemical processes. Some typical examples are thus:

cement manufacturing, iron and steel industries, oil and gas reservoirs, gas processing, oil refining, and ethylene production. Mitigation

associated with the capture of CO_2 from these industrial-based processes is small compared to that of the transportation and electricity
sectors; however, there may not be alternatives to the materials created
from these processes, such as cement, iron, and steel production, etc.

In the following, some of these aspects are delineated (briefly):

Cement manufacturing results in CO_2 emissions sourced from *calcination* in addition to the fuel combustion emissions of cement kilns. It
is estimated that the worldwide emissions from the cement industry
are approximately 2–4 Gt of CO_2 with approximately 52% and 48%
associated with the calcination process and cement kilns, respectively. Ordinary cement is a mixture of primarily di- and tricalcium
silicates ($2CaO \cdot SiO_2$, $3CaO \cdot SiO_2$) as well as small amounts of other
compounds:

[calcium sulfate ($CaSO_4$); magnesium, aluminum, and iron oxides;
and tri-calcium aluminate ($3CaO \cdot Al_2O_3$)].

The primary reaction that takes place in this process is the formation of
calcium oxide and CO_2 from calcium carbonate, which is highly endothermic and requires 3.5–6.0 GJ per ton of cement produced.

The steel-making industry is a multi-step technology. This industry
produces a combination of emissions, and the chemical processes constitute the estimated 1 Gt of CO_2 emitted worldwide. Steel making gen-
erates CO_2 as a result of carbon oxidation to carbon monoxide, which is
required for the reduction of hematite ore (Fe_2O_3) to molten iron (pig iron).
Another anthropogenic source of CO_2 is the combination of coal-burning
and limestone calcination. In the second stage of the steel-making process,
the carbon content of pig iron is reduced in an oxygen-fired furnace from
approximately 4–5% down to 0.1–1% and is known as the basic oxygen
steelmaking (BOS) process. Both these steps produce a steel-slag waste
high in lime and iron content.

In an oil refinery plant, crude oil, a mixture of various hydrocarbon
components ranging broadly in molecular weight, is fractionated from
lighter to heavier components. In a second stage, the heavier components are catalytically "cracked" to form shorter hydrocarbon chains. In
addition to producing CO_2 as a byproduct of the distillation and catalytic cracking processes, the heat and electricity required for the methane

reforming process used in H_2 production for hydrocracking and plant utilities produce additional CO_2.

Recovered natural gas from gas fields or other geologic sources often contains varying levels of non-hydrocarbon components such as CO_2, N_2, H_2S, and helium. Natural gas (primarily methane and ethane) and other light hydrocarbons such as propane and butane, to a lesser extent, are the valuable products in these cases, and often the generated CO_2 is a near-pure stream. Concentrations of approximately 20% CO_2 are not uncommon in large natural gas fields.

Exhaust Emissions: Comparing the sectors (electricity, transportation, industrial, commercial, and residential), currently the electricity sector is the largest emitter, representing 40% of total CO_2 emissions. Among all the sectors, comparing the different fossil fuel sources, that is, coal, petroleum, and natural gas, petroleum constitutes the majority of the emissions at approximately 43%. It is also suggested that CCS technologies are highly dependent upon the following four factors:

(1) the nature of the application, that is, a coal-fired power plant, an automobile, air, etc.;
(2) the concentration of CO_2 in the gas mixture;
(3) the chemical environment of CO_2, that is, the presence of water vapor, acid species (SO_2, NO_x), particulate matter (PM), etc.;
(4) the physical conditions of the CO_2 environment, that is, the temperature and pressure.

The concentration of CO_2 determines the energy in that the work required for separation decreases as the CO_2 concentration increases. The greater the CO_2 content (i.e., higher chemical potential) in a given gas mixture, the easier it is to carry out the separation (i.e., adsorption). If the CO_2 concentration in a gas mixture (such as flue gas, air, gas fuels) is too low, then one has to apply different separation techniques. For instance, in order for **membrane technologies** to be effective, a sufficient driving force (i.e., chemical difference across the membrane) is required. One of the many benefits of membrane technology arises from the act that it is low as regards capital cost. Membranes are a once-through technology in that the gas mixture enters the membrane in one stream and leaves the membrane as two streams, with one of the streams concentrated in CO_2. The chemical-process-based sources of CO_2 tend to have higher concentrations, making these processes targets for membrane technology application. Examples include ammonia, hydrogen, and ethanol production facilities. In addition, the chemical environment of CO_2 is important when considering the separation technology since some technologies may have a higher selectivity to other chemical species in the gas mixture.

For instance, one finds that in coal-fired flue gas, water vapor and acid gases (SO_2 and NO_x will also be present besides CO_2 for binding in solution. The effect of the temperature and pressure on the CO_2 capture is also of importance.

If a process occurs at high temperature or pressure it may be possible to take advantage of the work stored at the given conditions for use in the separation process. For instance, a catalytic reaction involving CO_2 may be enhanced at high temperature. It is important to mention that a catalytic approach for flue gas scrubbing is the case of NO_x reduction to water vapor and N_2 from the catalytic reaction of NO_x with ammonia across vanadia-based catalysts. This approach to NO_x reduction in a power plant is referred to as selective catalytic reduction. Non-catalytic NO_x reduction, in which ammonia is injected directly into the boiler, is also practiced, but it is not as effective as the catalytic approach. Membrane technology is another example, in that separation may be enhanced at high pressure.

The temperature balance of carbon capture has been analyzed. If the process is designed such that the CO_2 separation process (which effectively consumes the thermal energy) runs at the high temperature of the flue gas, then this would maximize the thermal content in the system. The flue gas temperature generally is around 650°C at the exit of the boiler, down to approximately 40–65°C at the stack. Current technologies such as absorption and adsorption are exothermic processes that are enhanced at low temperatures and in a traditional sense are not effective strategies for taking direct advantage of the thermal content of the high-temperature flue gas. For instance, the capture of CO_2 is most effective at low temperature for absorption and adsorption processes, with the regeneration of the solvent or sorbent most effective at elevated temperatures.

Membrane separation and catalytic-based technologies may, however, be enhanced at elevated temperatures (and pressures) available at exit boiler or gasifier conditions, depending on the specific technology.

Currently, the largest use of CO_2 is for EOR (enhanced oil recovery). This has been used for a few decades.

Compression and Transport of CO_2. Currently, on average, CO_2-EOR technology provides the equivalent of 5% of the U.S. crude oil production at approximately 280,000 barrels of oil per day. A limitation of reaching higher EOR production is the availability of CO_2. It is reported that some natural CO_2 fields can produce approximately 45 Mt CO_2 per year, with anthropogenic sources slowly increasing (currently 10 Mt CO_2 per year). It has also been suggested that the CO_2-EOR technology may also be a useful method to potentially store CO_2.

For instance, the CO_2 used for EOR is not completely recovered with the oil. In fact, only 20–40% of the CO_2 injected for EOR is produced with the oil, separated, and reinjected for additional production.

To date, EOR has not had any financial incentive to maximize CO_2 left below ground. In fact, since the cost of oil recovery is closely coupled to the purchase cost of CO_2, extensive reservoir design efforts have gone into minimizing the CO_2 required for enhanced recovery. If, however, the objective of CO_2 injection is to increase the amount of CO_2 left underground while recovering maximum oil, then the approach to the design question changes. If there were such an incentive, likely an even larger fraction would stay below ground, via modifications of EOR.

Another study was reported to have investigated the co-optimization of CO_2 storage with enhanced oil recovery. Their investigations concluded that traditional EOR techniques such as injecting CO_2 and water in a sequential fashion (i.e., a water-alternating-gas process) are not conducive to CO_2 storage. A modified approach includes a well-control process, in which wells producing large volumes of gas are closed and only allowed to open as reservoir pressure increases. In addition to co-optimization of CO_2 storage with EOR, ongoing efforts exist for coupled CO_2 storage with enhanced coal-bed methane recovery (ECBM) and potentially enhanced natural gas recovery from gas shales.

The post-combustion capture of CO_2 has taken place commercially for decades, primarily for the purification of gas streams other than combustion products.

It is important to recognize that usage of CO_2 in the food industry is not a mitigation option as the CO_2 is subsequently reemitted into the atmosphere, yet the usage of CO_2 continues to drive the advancement of separation technologies.

It is useful to consider the scale of CO_2 emissions associated with each of the primary energy resources. The annual emissions generated from coal, petroleum, and gas are on the order of 13, 11, and 6 Gt CO_2, respectively. Collectively, the emissions associated with the oxidation of fossil-based energy resources are on the order of 30 Gt CO_2 per year.

The largest chemicals produced worldwide are as follows: lime, sulfuric acid, ammonia, and ethylene production, on the order of 283, 200, 154, and 113 million tons in 2009, respectively.

(IEA, 2011).

4.2.2 Storage in Geological Rock Formations

In this geologic cross-section, supercritical CO_2 is stored underground in porous rock beneath a layer of impermeable shale. CO_2 does, in some cases, have an economic value, and currently one finds that research is being carried out on this aspect.

Some such economic aspects are carbonating beverages, usage in materials such as plastics or concrete, processing it to grow plants in enclosed greenhouses, processes that convert CO_2 into methane (CH_4) or liquid fuel.

In one large application of CCS, deep saline aquifers are considered to be main storage sites. These reservoirs typically lie some 2–4 km below the surface of the Earth, composed of 50-m-thick porous sandstone filled with saline water. In some parts of Earth, these geological formations are estimated to capable to store few thousands of billion tons of CO_2, sufficient for centuries of emissions. The CO_2 is expected to remain contained in such formations because they lie beneath impermeable shale, and capillary pressure in the sandstone pores holds the CO_2 in place. Over time, the brine reacts with the CO_2 to form solid calcium carbonate ($CaCO_3$).

4.2.3 Economical Aspects of CCS Technology Application

Although this subject is out of scope of this book, a few remarks about the CCS economy will be given. In general, pollution monitoring, control, and mitigation are expensive worldwide. Purification and control of drinking water is one of these examples. Regardless of cost, because it is health related, purification of drinking water has no constricting economic factor. Control and mitigation of all kinds is expensive. Another related issue, corrosion, is one of the biggest phenomena that costs billions worldwide. The monitoring and control of air is thus known for almost a century.

It is estimated that application of CCS to cement industry (IEAGHG, 2008), could increase the cost of production by 36% to 42% for oxy-fuel capture and 68% to 105% for post-combustion capture (IEAGHG, 2013c).

A reduction in the emissions of the cement could substantially reduce the embodied emissions in a building or house, with a relatively minor impact on the cost of the total construction.

- Fertilizers production: In the production of fertilizer, ammonia is the basis for nearly all synthetic fertilizers globally. With population growth, economic development and increasing competition for land use there will be increasing demand for productivity increases in food production.

  Presently, most ammonia is derived from hydrogen from fossil fuels, and inherent in the hydrogen production process is the stripping out of CO_2. It is suggested that with the application of CCS, the emissions from ammonia production could be reduced by 65% to 70%.
- Plastics and related industries: The cracking process is used to produce large amounts of ethylene. Ethylene is a building block for a wide variety of consumer products including plastics, polymers, and detergents. Ethylene is produced from cracking hydrocarbons, usually through steam. CO_2 is produced both through the generation of heat and from the cracking of hydrocarbons.

The chemical, steel, and cement industries are all known to produce CO_2 as a bi-product, and CCS is a major technology to achieve reductions in these carbon emissions. Emissions from industry accounted for around 26% of global CO_2 emissions (ca. 10 $GtCO_2$) each year. It is reported that besides other methods to reduce CO_2 emissions, currently CCS is the only viable technology available to achieve useful reductions (OECD/IEA, 2016).

CHAPTER 5

A Short Review of Different Carbon Dioxide (CO$_2$) Capture Processes (Adsorption on Solids and Absorption in Fluids)

In the capture of CO$_2$ gas, in relation to CCS, one finds many specific studies where different methods have been reported. It is useful to provide a short review of system studies that specifically apply to CO$_2$ adsorption/absorption capture in the current literature. The aim of this chapter is to provide the reader a more state-of-the-art information. There are two important procedures as reported in the literature that are of main interest (as described in Chapter 2.2 and 2.3):

adsorption of CO$_2$ on solids,

absorption of CO$_2$ in fluids.

Carbon capture technology is based on many different parameters. This means that one needs to apply the most optimum process at each

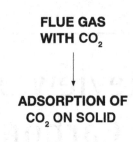

FIGURE 5.1 Schematic of gas adsorption and absorption (CCS) processes.

individual site. It is useful to describe some typical adsorption and absorption principles (surface chemistry based) from current literature (Figure 5.1). CO_2 is soluble in water. However, in aqueous solutions with different solutes (such as aqueous solutions of NaOH or amines), the solubility of CO_2 is considerably enhanced. The latter characteristic thus makes the process specific for separation of CO_2 from flue gases, with high purity.

However, besides these main CCS technologies, one finds some other related methods, including membrane gas separation, cryogenic CO_2 separation, CO_2 clathrate formation (Chapter 4; Appendix C).

In the current literature there are extensive studies on the adsorption data of CO_2 on solid surfaces, in relation to CCS application. Similarly, there are literature studies related to absorption of CO_2 in fluids (Chapter 5.2). It is also known that each application is specific to the particular project (based on different parameters).

5.1 ADSORPTION OF CO_2 (GAS) ON SOLIDS

In a system of a gas and a solid, one finds that under given conditions the gas may adsorb on the surface of the solid. The experimental methods used to study this process are different, depending on what parameter one needs to investigate. Gas adsorption on solids is studied by using different kinds of procedures, which may be (Appendix A):

amount of gas adsorbed/gram of solid,

volume of gas adsorbed/gram of solid,

heat evolved on adsorption,

spectroscopic analyses,

atomic microscopic analyses.

Gas adsorption technology is currently applied for CCS in some flue gas control (Figure 5.2). The flue gas is passed through a solid where adsorption (and subsequent desorption of CO_2) takes place.

This review is intended to provide useful information to the reader as regards the different research procedures being carried out in the literature. In the past decades many advances have been made on this subject.

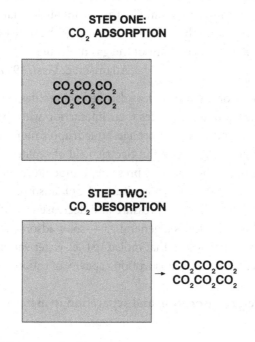

FIGURE 5.2   CCS process for flue gases.

DIFFERENT LITERATURE STUDIES ON ADSORPTION OF CARBON DIOXIDE (CO_2) ON SOLIDS

Carbon dioxide (CO_2) adsorption on zeolites: Principal factors

In this study a review was given, as regards advances in the principal structural and experimental characteristics of significance of the CO_2 adsorption on different zeolites (natural and synthetic sample) (Bonenfant et al., 2008). Experiments have indicated that the CO_2 adsorption is primarily related to a process where the inclusion of exchangeable cations (counter-cations) within the cavities of zeolites, which induce basicity and an electric field, two key parameters for CO_2 adsorption. The latter is related to the acidic characteristics of the CO_2 molecule. It was also found that these two parameters are different and related to the nature, distribution, and number of exchangeable cations.

The molecular structures of zeolite determine CO_2 adsorption characteristics:

> The effect of basicity and electric field in the cavities of zeolites was investigated. In fact, the basicity and electric field usually vary inversely with the Si/Al ratio in zeolites. Furthermore, the CO_2 adsorption will be expected to be limited by the size of pores (i.e., ratio of size of CO_2 to pore size) within zeolites and by the carbonate formation during the CO_2 chemisorption. The diffusion rate inside the pores will be dependent on the *Knudsen* constant of the gas molecule
>
> (Adamson & Gast, 1997; Birdi, 2017).

The polarity of the molecules adsorbed on zeolites represents a very important factor that influences their interaction with the electric field. The adsorbates that have the strongest quadrupole moment such as CO_2 might interact strongly with the electric field of zeolites, and this favors their adsorption. The effect of pressure, temperature, and presence of water was investigated. The magnitude of CO_2 adsorption was found to increase with the gas phase pressure and decrease with the rise of temperature. This behavior is a general gas—solid adsorption phenomena. Furthermore, it was found that moisture (i.e., water vapor) had a strong effect (decreasing) on the adsorption capacity of cationic zeolites.

Carbon dioxide gas adsorption and separation in metal-organic framework (MOF)

The need for controlling anthropogenic CO_2 emissions from various combustion processes (e.g., coal/oil/natural gas/wood burning) and lowering the emissions concentration of GHGs in the atmosphere has obvious procedure for environmental issues of mankind. Carbon capture and storage (CCS) is consider as the only viable process one for reducing/controlling the CO_2 emissions. In a recent study (Li et al., 2011) the adsorption of CO_2 on metal-organic frameworks (MOF) were analyzed. Even though a variety of technologies and methods have been developed, the separation of CO_2 from flue gas is still an important issue.

For example, metal-organic frameworks (MOFs) (crystalline porous materials) made by metal-containing nodes bonded to organic bridging ligands were studied. Researchers hoped that these MOFs were effective adsorbents or membrane materials in gas separation. A review on the research progress (from experimental results to molecular simulations) in MOFs for CO_2 adsorption, storage, and separations (adsorptive separation and membrane-based separation) that are directly related to CO_2 capture.

CARBON DIOXIDE CAPTURE BY ADSORPTION (REVIEW)

The basic aim of this investigation was to review different materials that can be useful as most effective adsorbents for CO_2 capture. In this study characteristics of different solid adsorbents were analyzed (Hinkov et al., 2016).

Carbon-based adsorbents for CO_2: It was reported that CO_2 may be recovered from flue gas by using non-reactive sorbents like carbonaceous porous solids and zeolites. High porous materials such as

activated carbons and charcoals exhibit CO_2 capture capacities ranging from 10 to 15% by weight. It was found that the selectivity is relatively low. Therefore, the CO_2 capture costs were estimated such that the carbon-based systems can be applicable when CO_2 purity is not more than 90%.

The adsorption on high-surface-area porous carbons synthesized by chemical activation (using petroleum coke as precursor and KOH as activation agent) were studied. The maximum CO_2 adsorption capture of 15.1 wt. % together with CO_2, and with a selectivity of 9.4 at 0.1 MPa were obtained for a sample activated at 973 K, which shows that it exhibits high effectivity.

Adsorption of CO_2 on carbon:

Adsorption on carbon was investigated (Gill et al., 1985). The procedure used was based on using phenol—formaldehyde resins + biomass residue + olive stones to prepare five different kinds of activated carbons for CO_2 separation at atmospheric pressure. These data were used to estimate the optimum values of activation temperature and burn-off degree that maximize CO_2 uptake by the activated carbons at 308 K and atmospheric pressure. The maximum adsorption capacity was found to be 9.3 wt. %.

In literature one finds different studies that are related to improving the adsorptive characteristics of porous adsorbents by modifying their surfaces (based on enhancing their basic feature). The latter is expected to enhance adsorption of acidic molecules (such as CO_2). These different modifications include exchange or substitution of cations within the metal framework, chemical treatment of the surface, or functionalization of the pores. The modifications usually result in a decrease of the surface area but increase the adsorptive selectivity and capacity for CO_2.

CO_2 Adsorption on MgO:

The adsorption equilibrium of CO_2 on chemically modified active carbons (by introducing MgO and S-CaO-MgO) have been studied (Yong & Rodrigues, 2001). The commercial super-activated carbons (MSC-30 and G08H with surface areas of 3370 m^2 g^{-1} and 2250 m^2 g^{-1}, respectively) were used as basic-activated carbon. The BET (Brunauer, Emmett, & Teller) surface area of the chemically modified carbon samples was found to be lower than those of the original carbon.

(Chapter 2)

These studies showed that it was not easy to enhance the adsorption capacity of CO_2 on carbon-based adsorbent at high temperature (573 K) and high pressure (0.1 MPa) by introducing a calcium oxide-magnesium oxide mixture only. However, the introduction of a hetero-element "S" and the increase of the polarity of carbon-based adsorbent was found to have a very favorable effect on the adsorption of CO_2 on carbon-based adsorbents at high temperature. The high-temperature ammonia treatment of commercialized carbon CWZ-35 activated carbon has been studied. The treatment was performed for 2 hours at elevated temperatures (from 473 to 1273 K). The solid surface was investigated by infrared spectroscopy (Fourier transform) (IRFT) of solid. It was found that in comparison with pristine activated carbon, all the carbons treated with ammonia showed an enhanced ability to adsorb CO_2.

Carbon Dioxide (CO_2) Adsorption on Mesoporous Aminopropylsilane-Functionalized Silica and Titania: (Microcalorimetry and in Situ Infrared Spectroscopy Studies)

CO_2 adsorption on two different calcined mesoporous supports, silica and titania, (which were functionalized with aminopropylsilane (APS)) (Knofel et al., 2009) has been investigated (enthalpy of adsorption). The solid samples were characterized using

infrared ATR (attenuated total reflectance) spectroscopy,
nitrogen sorption at 77 K,
and thermogravimetric analysis (TGA).

The functionalized silica and titania solids used were mesoporous and were grafted with 1.4 and 1.6 molecules of APS per nm^2, respectively. The infrared (ATR) analyses showed that the amine sites were changed and that the chemical properties of the support were significantly different. This shows that infrared (ATR) is a very useful method to analyze the structures of adsorbed species.

These studies showed that the NH_2^- bending vibration was shifted to lower wave numbers from 1597 to 1575 cm^{-1} from the silica- to the titania-grafted sample, respectively. This shift was attributed to different interactions with the surface hydroxyl groups of the silica and titania. The grafted samples were investigated for CO_2 adsorption by using both microcalorimetry and infrared (FTIR) spectroscopy.

The CO_2 adsorption properties were compared to the nongrafted support materials. The calorimeter (microcalorimetric) measurements indicated important enthalpies of adsorption at low CO_2 coverage (more than -80 kJ mol^{-1}) for the APS-grafted materials, indicating a strong reactivity between CO_2 and the amine sites. In situ infrared spectroscopy was used to study this reactivity. From these data it was concluded that this could be due to the formation of three different compounds at the solid surface (carbamate, carbamic acid, and bidentate carbonate).

Adsorption of Carbon Dioxide (on zeolitic imidazolate frameworks as selective CO_2 reservoirs):

In this study the adsorption of CO_2 on zeolites was investigated (Wang et al., 2008). In the case of CO_2 adsorbents, materials such as zeolitic imidazolate frameworks (ZIFs) have been found to exhibit porous crystalline tetrahedral networks that resemble those of zeolites: transition metals (Zn, Co) replace tetrahedrally coordinated atoms (for example, Si), and imidazolate links replace oxygen bridges. Further, an important characteristic property of these materials is that the structure adopted by a given ZIF is determined by link—link interactions.

As a result, systematic variations of linker substituents have yielded many different ZIFs that exhibit varying zeolite topologies. The materials are found to be chemically and thermally stable. The latter property is thus useful for adsorption applications. Two different kinds of porous ZIFs (ZIF-95 and ZIF-100) samples with structures have a scale and complexity previously unknown in zeolites.

Studies have shown that these materials have complex cage-like structures that may have 264 vertices (which are made up of 7,524 atoms). As expected from the adsorption selectivity recently documented for other similar types of materials, both ZIFs are found to selectively capture CO_2 from several different gas mixtures at room temperature, with ZIF-100 capable of absorbing 28 *liters per liter* of material at standard temperature and pressure. It was concluded, based on their high thermal and chemical stability and ease of fabrication, that ZIFs are useful solid materials for adsorption of CO_2 gas for applications in CCS ethnology.

Adsorption of Carbon Dioxide on Activated Carbon

Extensive data have been reported on the adsorption of CO_2 on activated carbon at different temperatures (Guo et al., 2006). The adsorption of CO_2 on a raw activated carbon **A** and three modified activated carbon samples **B**, **C**, and **D** at temperatures ranging from **303** to **333** K and the adsorption thermodynamics of these systems were analyzed.

The active ingredients impregnated in the carbon samples show significant influence on the adsorption for CO_2, and the volumes adsorbed on modified carbon samples **B**, **C**, and **D** are all larger than that on the raw carbon sample **A**.

However, the physical parameters such **as** surface area, pore volume, and micropore volume of carbon samples show no influence on the adsorbed amount of CO_2.

The data for samples A and B were found to be different than for C and D. The Freundlich equation (see equation) was found to give the best fit for adsorption on carbon samples **C** and **D**.

The isosteric heats of adsorption on carbon samples **A, B, C,** and **D** were estimated from the adsorption isotherm data (by applying the Clapeyron equation), which decreased with increasing surface adsorption. The values of the enthalpy of adsorption was reported to be in the range of **10.5 kJ/ mol** and 28.4 kJ/mol.

The carbon sample **D** exhibited the highest value, irrespective of the degree of adsorption. The observed entropy change associated with the adsorption for the carbon samples **A, B,** and **C** (above the surface coverage of **7** ml/gm) was lower than the theoretical value for mobile adsorption. However, it was higher than the theoretical value for mobile adsorption but lower than the theoretical value for localized adsorption for carbon sample **D.** Activated carbon is a highly microporous material with a large surface area and has been used as one of the main adsorbents for desulfurization. The method of removing CS_2 by using activated carbon at normal temperature was considered to be economical and appropriate.

In this adsorption study different active carbon solids (designated as **A, B, C,** and **D**) were used.

Sample **A:** This was a commercially activated carbon, **ZL-30,** which was obtained as a raw material (from China).

The three modified carbon samples, **B, C,** and **D,** were prepared by raw materials that were impregnated with different solutions. The saturated carbon samples were kept humid for 24 h at room temperature.

Sample **B:** This sample was impregnated with a 4% KOH solution and was heated at **373–393** K for **2–3** h and then dried in a vacuum oven at **353–373** K for **3** h.

Sample **C:** This sample was prepared by impregnating it with a mixture of ethylenediamine and ethanol (the volume ratio of ethylenediamine and ethanol was 2:l) and then dried at **343–363** K for **3** h.

Sample **D:** This sample was prepared by treating it with a mixture of 4% KOH, ethylene-di-amine, and ethanol (the volume ratio of ethylene-di-amine and ethanol was 2:l) and then drying it at **343–363** K for **3** h. The physical properties of the four carbon samples were also described.

The adsorption isotherms of CO_2 at different temperatures on the four activated carbon samples were studied. The adsorption data were analyzed by using different gas adsorption models (Yang, 1987; Dubinin, 1960).

The Freundlich adsorption equation was found to give the best fit for the adsorption on carbon sample **C** at **303, 313,** and **333** K and carbon sample **D** at **303, 313, 323,** and **333** K, respectively.

The Freundlich equation is given as

$$V = K P^{1/n} \tag{5.1}$$

where K and n are the Freundlich model constants. The magnitude of the quantity had isosteric heat of adsorption of CO_2 at a given specific surface coverage, V, which was calculated from the adsorption isotherms at different temperatures using the Clapeyron equation (Chapter 2):

$$h_{ad} = R T^2 (\partial \text{ Ln } P / \partial T)_{V,T}$$
$$= -R (\partial \text{ Ln } P / \partial (1 / T))_{V,T} \tag{5.2}$$

where R is the gas constant, T is temperature, and P is pressure. The best-fitting isotherm models were used to estimate the values of h_{ad}. The isosteric heats of adsorption for the four carbon samples at different surface coverage were analyzed. The value of adsorption heat on carbon sample **A** decreased slightly from 16.2 to 14.7 kJ/mol, and the surface coverage increased from **3** to 6 ml/gm. The isosteric heat of adsorption on carbon sample **B** was found to decrease from 12.3 to 10.5 kJ/mol, and the surface coverage increased from 4 to 8 ml/gm. The isosteric heat of adsorption on carbon sample **C** also showed a small decrease from 19.6 to 17.7 kJ/mol and an in- crease in the surface coverage from 5 to 9 ml/gm. In the case of sample **D**, a small decrease in the heat of adsorption was observed with increase of adsorption. The value of Q was measured to be 28 kJ/mol. At a surface adsorption of 6 mL/gm, the thermodynamic data for the different samples were analyzed (Choudhary & Mayadevi, 1996). The isosteric heats of adsorption for the four carbon samples at different surface loadings were analyzed. It was found that the value of adsorption heat on carbon sample **A** decreased slightly from 16.2 to 14.7 kJ/mol and the surface coverage increased from **3** to 6 ml/gm. The isosteric heat of adsorption on carbon sample B decreased from 12.3 to 10.5 kJ/mol and the surface coverage increased from 4 to 8 ml/gm. The isosteric heat of adsorption on carbon sample C showed only a small decrease from 19.6 to 17.7 kJ/mol and an increase in the surface coverage from 5 to 9 ml/gm. Further increase in adsorption exhibited a small decrease in the heat of adsorption on carbon sample D. The value of heat of adsorption was found to be 28 kJ/mol.

Adsorption of CO_2 on the modified surface of activated carbon:

A study on the influence of surface modification of activated carbon with gaseous ammonia on adsorption properties toward CO_2 was reviewed.

(Shafeeyan et al., 2014)

It was apparent from the literature survey that the surface chemistry of activated carbon strongly affects the adsorption capacity. In general, CO_2 adsorption capacity of activated carbon was found to be increased by the introduction of basic nitrogen functionalities into the carbon surface. This showed that CO_2 adsorption increased on basic carbon.

The impact of changes in surface chemistry and formation of specific surface groups on adsorption properties of activated carbon were studied. The solids were treated by different procedures: heat treatment and ammonia treatment (amination). These procedures were used to form activated carbon with basic surface and were compared. Amination was found to be suitable modification technique for obtaining efficient CO_2 adsorbents. These data thus show that CO_2 exhibits acidic properties.

Carbon Dioxide Adsorption Isotherms in Metal Organic Frameworks (Isotherms with Inflections and Steps)

As already mentioned, there is an adsorption mechanism in some systems such that gas molecules form multi-layer structures (Walton et al., 2008). The adsorption isotherm for CO_2 on MOFs showed inflections (as related to multilayer adsorption) that were found to be dependent on temperature. However, the adsorption isotherms for CO_2 in IRMOF-1 showed inflections that showed pronounced steps at lower temperatures.

These studies showed that the isotherm shapes can be predicted by molecular simulations using a rigid crystal structure, indicating that changes in the MOF crystal structure are not responsible for the steps in this system.

Adsorption characteristics of CO_2 on organically functionalized Amino-silanes:

Aminosilane-modified SBA-15 was prepared by grafting various aminosilanes on mesoporous silica SBA-15, and its adsorption characteristics toward CO_2 were examined.

(Hitoshi et al., 2)

The adsorption was carried out both in the presence and absence of moisture.

The data showed no effect on the amount of CO_2 adsorbed; the amounts were almost the same in the presence and in the absence of water vapor.

It was found that the efficiency of adsorption, defined as the number of adsorbed CO_2 molecules per nitrogen atom of aminosilane-modified SBA-15, increased with an increase in the surface density of amine.

The adsorbed species were analyzed by infrared spectroscopy.

These analyses showed that CO_2 was adsorbed on aminosilane-modified SBA-15 through the formation of alkylammonium carbamate in the presence and in the absence of water vapor.

It was suggested that amine pairs, on which CO_2 was adsorbed through formation of alkylammonium carbamate, increased with increasing surface density of amine.

The results were explained in relation to the influence of amine structure on the adsorption capacity.

Adsorption of CO_2 from air

The recovery of CO_2 directly from the atmosphere (with CO_2 concentration around 0.04%: 400 ppm) has been investigated by some reports in literature (Sanz-Pérez et al., 2016). In these studies, different procedures were reported and compared with other methods.

Carbon Dioxide Adsorption (Metal Substitution) in a Coordination Polymer with Cylindrical Pores

A series of four different iso-structural microporous coordination polymers (MCPs) differing in metal composition were investigated for CO_2 adsorption. It was found that these solids had adsorption characteristics including enhanced adsorption of CO_2 at low pressures and temperature (Casket et al., 2008). These conditions are particularly relevant for capture of flue gas from coal-fired power plants. These conditions also keep costs low.

For example, for CO_2 at 0.1 atm/296 K, the degree of adsorption was found to be about 23.6%.

A magnesium-based material is presented that has the highest surface area magnesium MCP yet reported and displays ultrahigh affinity based on heat of adsorption for CO_2.

From these studies it was shown that physisorptive materials exhibit affinities and capacities competitive with amine sorbents while greatly reducing the energy cost associated with regeneration.

Adsorption separation of CO_2 from flue gas of natural gas-fired boiler by a novel nanoporous "molecular basket" adsorbent

In all flue gases CO_2 is found in mixture with other gases (such as CO, NOx, SO_2, etc.). It is thus useful to study the adsorption characteristic of CO_2 in the presence of different gases (as found in flue gas).

In a recent study a novel nano-porous CO_2 (a so-called molecular basket) adsorbent was developed and applied in the separation of CO_2 from the flue gas of a natural gas-fired boiler (Xu et al., 2005a, b). The nanoporous CO_2 "molecular basket" adsorbent was prepared by uniformly dispersing polyethylenimine (PEI) into the pores of mesoporous molecular sieve MCM-41. These studies showed that use of MCM-41 and PEI had a synergetic effect on the CO_2 adsorption. The rates of CO_2 adsorption/desorption of PEI were also greatly improved. Adsorption separation results showed that CO_2 was selectively separated from simulated flue gas and flue gas of a natural gas-fired boiler by using this novel adsorbent.

The adsorbent was found to adsorb a very small quantity of N_2, SO_2, and CO in the flue gas. Furthermore, the moisture had a promoting effect on the adsorption separation of CO_2 from flue gas. The adsorbent simultaneously adsorbed CO_2 and NO_x from flue gas. The adsorbed amount of CO_2 was around 3000 times higher than that of NO_x. The adsorbent was stable in several cyclic adsorption/desorption operations. However, very little NO_x was desorbed after adsorption, indicating the need for pre-removal of NO_x from flue gas before capture of CO_2 by this novel adsorbent.

Role of Amine Structure on CO_2 Adsorption from Ultra-dilute Gas Streams such as Ambient Air

Investigations on the adsorption properties of primary, secondary, and tertiary amine materials are used to evaluate what amine type(s) are best suited for ultra-dilute CO_2 capture applications (Didas et al., 2012). A series of comparable materials comprised of primary, secondary, or tertiary amines ligated to a mesoporous silica support via a propyl linker are used to systematically assess the role of amine type. Both CO_2 and water adsorption isotherms were obtained for these materials in the range relevant to CO_2 capture from ambient air, and it is demonstrated that primary amines are the best candidates for CO_2 capture from air. It was found that primary amines possess both the highest amine efficiency for CO_2 adsorption as well as enhanced water affinity compared to other amine types or the bare silica support. These studies showed that the rational design of amine adsorbents for the extraction of CO_2 from ambient air should focus on adsorbents rich in primary amines.

Advances in principal factors influencing carbon dioxide adsorption on zeolites

Adsorption is reported to be principally governed by the inclusion of exchangeable cations (counter-cations) within the cavities of zeolites, the nature, distribution fact, and the ratio of molecules adsorbed on zeolites represent very important factors that influence their interaction with the electric field (Bonenfant et al., 2008). High quadrupole moments in adsorbates, such as a field of zeolites, favor their adsorption. The pressure, temperature, and presence of water seem to be the most important experimental conditions that influence the gas phase pressure, which decreases with the rise of temperature. The presence of water significantly decreases adsorption capacity of cationic zeolites by decreasing strength and heterogeneity of the electric field and by favoring the formation of bicarbonates. It was concluded that the zeolites structural characteristics could be changed, thus giving enhanced adsorbents that could be applied for capturing the industrial emissions of CO_2.

Highly efficient separation of carbon dioxide by a metal-organic framework with open metal sites

Selective capture of CO_2, which is essential for natural gas purification and CO_2 sequestration, has been reported in zeolites, porous membranes, and amine solutions

(Britt et al., 2009).

However, all such systems require substantial energy input for release of captured CO_2, leading to low energy efficiency and high cost. A new class of materials named metal-organic frameworks (MOFs) has also been demonstrated to take up voluminous amounts of CO_2. However, these studies have been largely limited to equilibrium uptake measurements, which are a poor predictor of separation ability, rather than the more industrially relevant kinetic (dynamic) capacity. Here, we report that a known MOF, Mg-MOF-74, with open magnesium sites, rivals competitive materials in CO$_2$ capture, with 8.9 wt. % dynamic capacity, and undergoes facile CO$_2$ release at significantly lower temperature, 80°C. Mg-MOF-74 offers an excellent balance between dynamic capacity and regeneration. These results demonstrate the potential of MOFs with open metal sites as efficient CO$_2$ capture media.

Adsorption separation of carbon dioxide from flue gas of natural gas-fired boiler (nanoporous "molecular basket" adsorbent)

A novel nano-porous CO$_2$ "molecular basket" adsorbent was developed and applied in the separation of CO$_2$ from the flue gas of a natural gas fired boiler (Xu et al., 2005a, b). In all flue gases one finds other gas pollutants (SO, CO, etc.), besides CO$_2$. Therefore it is important that the adsorbents can be selective in adsorption property. The nano-porous CO$_2$ "molecular basket" adsorbent was prepared by uniformly dispersing poly-ethylenimine (PEI) into the pores of mesoporous molecular sieve MCM-41. The use of MCM-41 and PEI had a synergetic effect on the CO$_2$ adsorption. The rates of CO$_2$adsorption/desorption of PEI were also greatly improved. Adsorption separation results showed that CO$_2$ was selectively separated from simulated flue gas and flue gas of a natural gas-fired boiler by using this novel adsorbent. The adsorbent adsorbed very little N$_2$, O$_2$ and CO in the flue gas. It was also found that moisture had a promoting effect on the adsorption separation of CO$_2$ from flue gas. The adsorbent was found to simultaneously adsorbed CO$_2$ and NO$_x$ from flue gas. However, the adsorbed amount of CO$_2$ was around 3000 times larger than that of NO$_x$. The adsorbent was stable in several cyclic adsorption/desorption operations. However, very little NO$_x$ desorbed after adsorption indicating the need for pre-removal of NO$_x$ from flue gas before capture of CO$_2$ by this novel adsorbent.

Microporous metal-organic framework (MOF) with potential for carbon dioxide capture at ambient conditions

Carbon dioxide capture and separation are important industrial processes that allow the use of carbon dioxide for the production of a range of chemical products and materials and to minimize the effects of carbon dioxide emission. Porous metal-organic frameworks are promising materials to achieve such separations and to replace current technologies, which use aqueous solvents to chemically absorb carbon dioxide.

(Xiang et al., 2012)

These studies showed that a metal-organic framework (UTSA-16) displays high uptake (160 cm³ cm⁻³) of CO_2 at ambient conditions, making it a potentially useful adsorbent material for post-combustion CO_2 capture and biogas stream purification. This has been further confirmed by simulated breakthrough experiments. The high storage capacities and selectivities of UTSA-16 for CO_2 capture are attributed to the optimal pore cages and the strong binding sites to CO_2, which have been demonstrated by neutron diffraction studies.

ADSORPTION OF GASES ON SOLID CARBON (POROUS): EFFECT OF PORE SIZE

The concentration of a gas adsorption is the process by which a density increase is seen in the interfacial layer in a system containing a solid sorbent and a gaseous sorbate. Internal adsorption occurs on the walls of openings in the sorbent that are deeper than they are wide. These internal openings are referred to as pores, and pores are classified according to their width (Sing et al., 1985; Mosher, 2011).

A pore (in general) is an opening in the sorbent material that is deeper than it is wide and that is accessible to the adsorbing gas. For pores of adsorption, however, when considering a system that has reached equilibrium with a gas pressure equally distributed over all of the surface area inside the pore network, pore size is an adequate descriptor for the system.

The primary area of interest when considering adsorption at any single point is the distance to the opposite pore wall. All the pore volumes where the walls are of a similar distance across can be grouped together without regard to location. As a consideration of geometry, a block of sorbent with many pores of different widths whose pores are a homogeneous width through the whole sample will have the same adsorption characteristics as a block with pores erratic in every way as long as the aggregate pore size distribution is the same over the two samples. It will be also expected that the different pore sizes (and shape) can also be expected to exhibit different behavior during adsorption as pressure increases.

For example: Micropores will experience primarily physisorption at the pore-filling stage only, whereas meso- and macropores will exhibit two different stages. The first stage with single and multilayer adsorption, followed by capillary condensation (Sing et al., 1985). There are several methods for analyzing the data. The adsorption of gas on solid is measured as

. . . gas density versus pressure.

It is important to note that simulations and calculations of these densities are actually calculated from fugacity rather than simply pressure, where fugacity is the deviation of the gas pressure behavior from that of an ideal gas (Chattoraj & Birdi, 1984; Myers, 1983; Myers & Prausnitz, 1965; Myers & Monson, 2002).

The gas adsorption on a solid is defined as follows. "Total" adsorption is defined as the total quantity of gas that can be found in the pore space. Thus, this quantity (of adsorption) consists of both the gas in the bulk as well as the gas that is adsorbed at the solid surface. In general, this procedure of separating two quantities, that is, the adsorbed phase from the gas phase, is complicated. Another quantity, "absolute" adsorption, is defined as the amount of gas present only in the adsorbed state. Furthermore, one defines the "excess" adsorption, which is equivalent to absolute adsorption for pressure regimes that are not extreme. This procedure is analogous to the analyses of surface adsorption (excess) in solutions (Chattoraj & Birdi, 1984; Myers & Prausnitz, 1965; Myers, 1983, 1989; Myers & Clever, 1974; Myers & Monson, 2002; Myers & Sircar, 1972; Myers, 2005).

Carbon Dioxide Adsorption and Desorption on Malaysian Coals

The adsorption desorption of CO_2 on coal samples was reported (Abunowara et al., 2016). The data were analyzed by using different gas—solid adsorption theories.

The Langmuir isotherm model is based on the assumption that gas adsorption occurs at specific homogenous sites within the adsorbent surface. It explains that the adsorbent has a finite capacity for the adsorbate. The data was analyzed by using the following relation between the gas adsorbed and the pressure:

$$M_{ads} = \left(n_m b\, p_{gas}\right)/\left(1 + b\, p_e\right) \tag{5.3}$$

where M_{ads} is the amount of CO_2 adsorbed (mmol/gram) of coal, p_{gas} is the equilibrium pressure (bar), n_m is the maximum adsorption for the solid, and b is the constant (Langmuir, 1918).

In summary: the analyses of four coal samples from different locations were tested for CO_2 adsorption/desorption at temperatures of 273 K and 298 K and pressure up to 1 bar. The S3 has the highest adsorption capacity by 0.73 mmol/g. According to IUPAC classification of adsorption isotherms, CO_2 adsorption isotherms of all coal samples

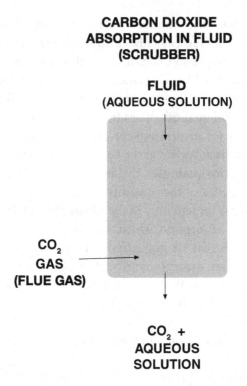

**CARBON DIOXIDE
ABSORPTION IN FLUID
(SCRUBBER)**

**FLUID
(AQUEOUS SOLUTION)**

**CO₂
GAS
(FLUE GAS)**

**CO₂ +
AQUEOUS
SOLUTION**

FIGURE 5.3   A typical gas scrubber for gas (flue gas) absorption.

follow the type I isotherm (Figure 2.7), which most probably describes adsorption limited to a few molecular layers (micropores). The results of adsorption and desorption isotherms demonstrate that there is hysteresis between adsorption and desorption isotherms for all coal samples. The coal samples S1 and S2 have the highest hysteresis level compared to coal samples S3 and S4, which show a positive hysteresis between their adsorption and desorption isotherms. According to hysteresis classifications, the hysteresis during the CO_2 adsorption and desorption process for all coal samples type H3 which describes the shape of pores. The analyses of data showed that the material exhibits heterogeneous surface properties.

5.2 ABSORPTION OF GASES (GAS ABSORPTION REVIEW)

The absorption of CO_2 in different fluids has been investigated extensively in the current literature. There are two kinds of approaches to this method. In one case, one uses aqueous solutions of amines or other similar basic molecules (CO_2 is an acidic molecule). In another approach, the fluid is a mixture of an aqueous solution with a non-miscible additive (such as an organic water-insoluble component).

Another CO_2 capture method is the **absorption** process (Benadda et al., 1994; Bruining et al., 1986; Danckwerts & Sharma, 1966; Eckenfelder, 1961; Joosten & Danckwerts, 1972; Cents et al., 2001), in which a gas is captured in a fluid (generally a solution). The gas is bubbled through a suitable apparatus with maximum contact between the gas and the fluid (Figure 5.2).

ABSORPTION PROCESS:

GAS . . . . BUBBLES . . . . FLUID. (CAPTURES CO2)

In this section some studies are reported that examine absorption of CO_2 in different fluids in relation to CCS.

For example, CO_2, when bubbled in a solution of NaOH solution, is known to form sodium carbonate (Na_2CO_3,.). This shows that CO_2 in water behaves as a weak acidic molecule and readily interacts with bases. In the following, some typical studies on the absorption of CO_2 in solutions is delineated. This is useful for various reasons. It shows primarily the scope of absorption process. It also indicates the various physical aspects of gas absorption phenomena.

Carbon Dioxide Postcombustion Capture: A Novel Screening Study of the Carbon Dioxide Absorption Performance of 76 Amines (in aqueous solutions)

The CO_2 absorption capacities of 76 different amines were studied (Puxty et al., 2009). The significant and rapid reduction of greenhouse gas emissions is recognized as necessary to mitigate the potential climate effects from global warming. The post-combustion capture (PCC) and storage of carbon dioxide (CO_2) produced from the use of fossil fuels for electricity generation is a key technology needed to achieve these reductions. The most mature technology for CO_2 capture is reversible chemical absorption into an aqueous amine solution. In this study (Puxty et al., 2009), the results from measurements of the CO_2 absorption capacity of aqueous amine solutions for 76 different amines are presented. Measurements were made using both a novel isothermal gravimetric analysis (IGA) method and a traditional absorption apparatus. Seven amines, consisting of one primary, three secondary, and three tertiary amines, were identified as exhibiting outstanding absorption capacities.

These amines were identified to exhibit some structural features: steric hindrance and hydroxyl functionality of 2 or 3 carbons from the nitrogen.

Initial CO_2 absorption rate data from the IGA measurements were also used to estimate relative absorption rates. Most of the outstanding performers in terms of capacity also showed initial absorption rates comparable to the industry standard mono-ethanolamine (MEA). From these studies it was concluded that such adsorbents are useful materials for gas absorption (both as regards capacity and kinetics).

Kinetics and modeling of carbon dioxide absorption into aqueous solutions of N-methyldiethanolamine

In a recent study, a wetted-sphere absorber was used to investigate kinetic data for the aqueous phase reaction between carbon dioxide (CO_2) and N-methyldiethanolamine (MDEA) (Rinker et al., 1995). Experiments were carried out over the temperature range of 293–342 K for partial pressures of CO_2 near atmospheric pressure and for 10–30 mass% MDEA. The data were found to be consistent with a mechanism in which MDEA catalyzes the hydrolysis of CO_2. Three different mathematical models that are based on Higbie's penetration theory were developed and used to estimate the forward rate coefficient of the MDEA-catalyzed hydrolysis of CO_2 reaction. A comparison of the numerical results of the three models indicated that the effect of the reaction between hydroxide and CO_2 is significant, especially when estimating the rate coefficient of the CO_2 / MDEA reaction for unloaded aqueous MDEA solutions. If one neglects the CO_2 / OH^- reaction, this may lead to large errors in the rate coefficient for the MDEA-catalyzed hydrolysis reaction.

The second-order rate coefficients of the MDEA-catalyzed hydrolysis reaction for 10% MDEA solution were estimated according to the most general model (by the Arrhenius type equation):

$$k_{21} = 2.91 \times 10^7 \exp(-4579\ T) \tag{5.4}$$

Carbon dioxide absorption with aqueous potassium carbonate promoted by piperazine

In general, various commercial processes for the removal of carbon dioxide from high-pressure gases apply aqueous potassium carbonate systems enhanced by secondary amines. A thermodynamic and kinetic study for aqueous potassium carbonate enhanced by the addition of piperazine was reported.

(Tim et al., 2004)

Investigations were carried out at typical absorber conditions for the absorption of CO$_2$ from flue gas.

In these studies piperazine used as an additive in 20–30 wt% potassium carbonate, was investigated in a wetted-wall column using a concentration of 0.6 m at 40–80°C. The addition of 0.6 m piperazine to a 20 wt% potassium carbonate system decreased the CO$_2$ equilibrium partial pressure by approximately 85% at intermediate CO$_2$ concentration.

The degree of distribution of piperazine species in the solution was determined by proton NMR. Using the speciation data and relevant equilibrium constants, a model was developed to predict system speciation and equilibrium parameters. The addition of 0.6 m piperazine to 20 wt% potassium carbonate gave an increase in the rate of CO$_2$ absorption by an order of magnitude at 60°C. The rate of CO$_2$ absorption in solution was the same as in the solution of 5.0 M MEA. The addition of 0.6 m piperazine to 20 wt% potassium carbonate also increased the heat of absorption from 3.7 to 10 kcal / mol. The degree of capacity of CO$_2$ was found to be from 0.4 to 0.8 mol-CO2 /kg-H$_2$O, for PZ/K2CO3 solutions (which was same as for other amines).

Absorption of carbon dioxide into aqueous piperazine: reaction kinetics, mass transfer and solubility

The absorption of carbon dioxide into aqueous solutions of piperazine in a wetted wall container (Bishnoi & Rochelle, 2000) was studied, from 298 to 333 K in solutions of 0.6 and 0.2 M aqueous piperazine. It was found that the apparent reaction rate is first order in both carbon dioxide and piperazine with a value of 53,700 m^3/kmol s at 25°C. The apparent second-order rate constant follows an Arrhenius temperature dependence over the range studied with an activation energy of 3.36×104 kJ/ kmol. Solubility in 0.6 MPZ was measured by bracketing absorption and desorption in the wetted wall contactor at 313 and 343 K.

The data of the phase equilibrium were modeled by considering the following piperazine species: piperazine carbamate, di-carbamate, protonated carbamate, and protonated piperazine. Henry's law and the dissociation of CO_2 to form bicarbonate and carbonate were also considered. The carbamate stability constant and pKa for piperazine carbamate were regressed from the VLE data. Although shown to be present by NMR, the dicarbamate is not the dominant reaction product at any loading.

The carbamate stability constant is comparable to other secondary amines such as di-ethanolamine (DEA), but the apparent second-order rate constant is an order of magnitude higher than primary amines such as monoethanolamine (MEA) or di-glycolamine (DGA). The second-order rate constant obtained in this work is much higher than previously published values for the piperazine/CO_2 reaction. These previous studies were limited by mass transfer limitation of products and reactants and were not a true measurement of the kinetics of CO_2/piperazine. There is some evidence that the reactivity of piperazine is due to its cyclic and diamine characteristics.

Carbon dioxide recovery from post-combustion processes: Can gas permeation membranes compete with the absorption process?

In a recent study (Favre, 2007) the gas permeation processes based on dense polymeric membranes were reported. Various studies indicate that absorption processes (using packed towers or membrane contactors) are the preferred technology. A critical analysis of dense polymeric membrane capture processes versus amine absorption when applied after combustion (i.e., flue gas treatment) was reported. Technological and scientific challenges, as well as prospects for future developments are discussed. The potential of dense polymeric membranes to solve the flue gas treatment problem may have been underestimated.

Kinetics of the reactive absorption of carbon dioxide in high CO_2-loaded, concentrated aqueous monoethanolamine solutions

The kinetics of the reaction between carbon dioxide and high CO_2-loaded, concentrated aqueous solutions of monoethanolamine (MEA) were investigated over the temperature range from 293 to 313K (Aboudheir, 2003). The concentration of MEA was varied from 3 to 9M, and CO_2 loading from ~0.1 to 0.49 mol/mol. The experimental kinetic data were obtained in a laminar jet absorber at various contact times between gas and liquid. The obtained experimental data were interpreted with the aid of a numerically solved absorption-rate/ kinetic model. Based on the data the results were used to develop a new molecular kinetics model for CO$_2$ reaction with MEA solutions. These model studies were compared to kinetic models in the literature. The model was comprehensive enough to describe for the first time the absorption of CO$_2$ in highly concentrated and high CO$_2$-loaded aqueous MEA solutions for a wide temperature range.

Kinetics of Carbon Dioxide Absorption into Aqueous Potassium Carbonate and Piperazine Solutions

The absorption rate of CO$_2$ was measured in a wetted-wall column. The absorptions of CO$_2$ in solutions of 0.45–3.6 m piperazine (PZ) and 0.0–3.1 m potassium carbonate (K$_2$CO$_3$) at 25–110°C (Cullinane & Rochelle, 2006) were investigated. The data were analyzed with a kinetic model.

It was found that the model indicated the absorption rate approaches second-order behavior with PZ and is highly dependent on other strong bases. In 1 M PZ, the overall rate constant is 102,000 s$^-$$_1$, which is found to be 20 times higher than in mono-ethanolamine solutions. The magnitude of the activation energy was found to be of the order of 35 kJ/mol, similar to other amine–CO$_2$ reactions. Rate constants for contributions of carbonate, PZ carbamate, and water to the rate were determined according to base catalysis theory. The addition of neutral salts to aqueous PZ increases the apparent rate constant. In 2.7 M NaCl + 0.6 M PZ, the overall rate constant is increased by a factor of 7. Ionic strength effects were accounted for within the rigorous model of K$^+$/PZ mixtures. The absorption rate in concentrated K$^+$/PZ mixtures is up to 3 times faster than in 30 wt% mono-ethanolamine. At high pressures, the reaction was found to be instantaneous but was still influenced by reaction kinetics.

Kinetics and Modeling of Carbon Dioxide Absorption into Aqueous Solutions of Diethanolamine

The kinetics of the reaction between CO$_2$ and aqueous di-ethanolamine (DEA) were estimated over the temperature range of 293–343 K from absorption data obtained in a laminar-liquid jet absorber (Rinker et al., 1996). The absorption data were obtained over a wide range of DEA concentrations and for CO$_2$ partial pressures near atmospheric. A rigorous numerical mass-transfer model based on penetration theory in which all chemical reactions are considered to be reversible was developed and used to estimate kinetic rate coefficients from the experimental absorption data. The kinetic data were found to be consistent with the zwitterion mechanism. The scarce zwitterion rate coefficient estimates reported in the literature are in fair agreement with the results of these data.

Carbon dioxide absorption in aqueous mono-ethanolamine solutions

Experiments have been carried out on the absorption of carbon dioxide in mono-ethanolamine (MEA) aqueous solutions (Astarita, 1961). In these studies, a laminar liquid jet absorber was used. From these studies, the results obtained (which covered the contact time range between 0.02 and 0.07 sec) showed good agreement with the penetration theory equations as well as with the kinetics of the mechanism of the chemical reaction that presumably takes place in the liquid phase. In these studies, a disk column absorber was used. The results were found to be reproducible and have been correlated in compact form with an empirical equation.

Investigation of amine amino acid salts (in aqueous media) for carbon dioxide absorption

The carbon dioxide absorption by aqueous solutions of amine amino acid salts (AAAS) formed by mixing equi-normal amounts of amino acids, for example, glycine, β-alanine, and sarcosine, with an organic base, 3-(methyl-amino) propylamine (MAPA) was investigated by comparison with monoethanolamine (MEA) and with amino acid salt (AAS) from amino acid neutralized with an inorganic base, potassium hydroxide (KOH).

(Aronu et al., 2010)

CO_2 absorption and desorption experiments were carried out on the solvent systems at 40°C and 80°C, respectively. Experimental results showed that amine amino acid salts have similar CO_2 absorption properties to MEA of the same concentration. They also showed good signs of stability during the experiments. Amino acid salt from an inorganic base, KOH, showed poorer performance in CO_2 absorption than the amine amino acid salts (AAAS) mainly owing to a lower equilibrium temperature sensitivity. AAAS showed better performance than MEA of the same concentration. AAAS from the neutralization of sarcosine with MAPA showed the best performance and the performance could be further enhanced when promoted with excess MAPA. The solvent comparison is semi-quantitative because of the bubble structure, and thus the gas—liquid interfacial area may not be the same for all experiments; however, superficial gas velocities were kept constant.

Effect of Steric Hindrance on Carbon Dioxide Absorption into New Amine Solutions: Thermodynamic and Spectroscopic Verification through Solubility and NMR Analysis

Currently it is obvious that efficient carbon capturing processes are needed for the treatment of flue gases (especially CO_2). The solubility and absorption of CO_2 in different solutions was investigated (Park et al., 2003). The absorbances of CO_2 in aqueous solutions of 2-amino-2-hydroxymethyl-1,3-propanediol (AHPD), a sterically hindered amine, were investigated. The latter were examined as a potential CO_2 absorbent and compared with the most commonly used absorbent, mono-ethanolamine (MEA) solution, through equilibrium solubility measurements and ^{13}C NMR spectroscopic analyses (Park et al., 2003). The solubilities of CO_2 in aqueous 10% AHPD solutions were found to be higher than those in aqueous 10% MEA solutions above 4 kPa at 298.15 K. However, below 4 kPa, the solubility behavior appeared to be the opposite. The solubility difference between these two solutions increased with CO_2 partial pressures above the crossover pressure. Equilibrated CO_2–MEA–H_2O and CO_2–AHPD–H_2O solutions at various CO_2 partial pressures ranging from 0.01 to 3000 kPa were analyzed by ^{13}C NMR spectroscopy to provide a more microscopic understanding of the reaction mechanisms in the two solutions. In the CO_2–amine–H_2O solutions, amine reacted with CO_2 to form mainly the protonated amine (AMH^+), bicarbonate ion (HCO_3^-), and carbamate anion ($AMCO_2^-$), where the quantitative ratio of bicarbonate ion to carbamate anion strongly influenced the CO_2 absorption in the amine solutions.

A profusion of bicarbonate ions, but a very small amount of carbamate anions, were identified in the CO_2–AHPD–H_2O solution, whereas a considerable amount of carbamate anions was formed in the CO_2–MEA–H_2O solution. The AHPD molecule contains more hydroxyl groups than the nonhindered molecule MEA. This was related to the chemical shifts in its ^{13}C NMR spectra. In contrast, MEA appeared to be insensitive to pH. The data showed strong interrelations among CO_2 solubility, CO_2 partial pressure, bulkiness of the amine structure, and pH for these organic absorbents.

CARBON DIOXIDE SEPARATION FROM FLUE GAS USING ABSORPTION IN AMINE-BASED SYSTEMS

A major source of CO_2 is the combustion of fossil fuels (coal, oil, and gas) used to supply energy in different forms, such as heat, electricity, mechanical work, etc. Global emissions of CO_2 from fossil fuel combustion increased from 20.7 billion tones (Gt) in 1990 to 24.1 Gt in 2002 (i.e., an increase of 16% in comparison to 1990) (Metz et al., 2005). In order to mitigate this increase of CO_2 in air, one needs to capture the gas, as by CCS. It is argued that besides the cost needed for CCS application, CCS is highly useful for control of the increasing CO_2 and its effect on climate. Actually, CO_2 capture is already an industrial technology, used today to process natural gas. It is commonly applied in the manufacture of fertilizers, in the food-processing industry, and in the energy sector (the oil and gas industry). The main problem is generally the low concentration of CO_2 in the flue gas.

In this study the flue gas consisted of the pollutants SO_2 and CO_2. Hence, SO_2 was first absorbed in the flue gas desulphurization (FGD) process (Ionel et al. (2008). The aqueous solutions were made up of sodium hydroxide or calcium oxide dissolved in water, forming a strongly alkaline solution; the mass ratio of NaOH to H_2O was 1:100, respectively, whereas the molar ratio of Ca:S was 1.5:1. In the FGD unit, the flue gases react with the selected aqueous solutions, which are pumped to the scrubber with a flow rate of 2 l/h. Sulfur dioxide (SO_2) reacts with sodium hydroxide and forms a mixture of sodium sulfite (Na_2SO_3) and water. After that, sodium sulfite additionally absorbs sulfur dioxide, resulting in a chemical compound of sodium hydrogen sulfite ($NaHSO_3$). There is also another reaction: a solution of sodium hydroxide reacts with other acid gases. For example, CO_2 reacts with aqueous sodium hydroxide to form sodium carbonate (Na_2CO3). When injecting calcium hydroxide as an additive in the FGD, solid calcium carbonate is formed. The following absorption processes are present:

$$4NaOH + 2SO_2 + O_2 = Na_2SO_4 + H_2O \tag{5.5}$$

And:

$$Ca(OH)_2 + SO_2 = Ca\,SO_4 + H_2O \tag{5.6}$$

After the removal of sulfur dioxide, the flue gas is pumped to the CO_2 absorption unit. The CO_2 is removed from the flue gas by means of a chemical solvent. An aqueous solution of mono-ethanolamine was selected as the solvent, with concentration around 35%. The main reaction between CO_2, a weak acid, and mono-ethanolamine, a weak base, is found to be reversible. Under these circumstances, if aqueous MEA is cooled to the temperature levels of 40–60°C, then the chemical solvent retains the CO_2.

However, when the MEA solution is heated up to 120–140°C, it releases the CO_2 gas, and the regeneration of the chemical solvent takes place.

The following chemical reactions occur when CO_2 is absorbed by MEA (Junkrer & Folmer, 1998; Kohl & Nielson, 1997; Petrova et al., 2006; Ionel et al., 2008):

$$2HOCH_2H_4NH_2 + H_2O + CO_2 ==== HOC_2H_4NCO_2 \\ + HOC_2H_4NH_3+ \tag{5.6}$$

$$2HOC_2H_4NH_2 + H_2O + CO_2 ==== 2HOC_2H_4NH_3+ + CO_3^{-2}+ \tag{5.7}$$

$$HOC_2H_4NH_2 + H_2O + CO_2 ==== HOC_2H_4NH_3 + + HCO_3^- \tag{5.8}$$

The absorption reaction was found to be driven from left to right. CO_2 is recovered and cooled to 40°C. In these studies, it was found that about 66% CO_2 was absorbed in a solution of 35% MEA.

During the absorption process, the reactions proceed from left to right.

In the scrubber, the solution from the bottom of the column ("rich" MEA solution), containing the chemically bound CO_2, was passed through a cooler. Afterwards, it was pumped to the desorber where it is heated in counter flow up to 120–140°C by the flue gas stream. However, during regeneration, the reactions proceed from right to left; thus, CO_2 and H_2O evolve separately from the amine solution. The captured CO_2 (99%) is captured through the top of the separation device. It can then be compressed and stored. The "lean" solution of MEA, containing far less CO_2, is cooled down to 40°C in a cooler, and recycled back to the absorber, for further additional CO_2 capture and continuity of the global process.

The effect of temperature inside the combustor were investigated. In some experiments the temperature reached as high as 980°C (for few seconds). The decrease of the gas temperature was caused by the fuel-feeding interruption. The concentration of SO_2 in the flue gas before and after the scrubber was treated with an alkaline solution of 1 wt% and 2 wt% NaOH. The concentration of SO_2 was in the range of 50–75 ppm after the scrubber. CO_2 absorption into 35 wt% MEA was relatively high with an absorption efficiency of 66% (Metz et al., 2005). The reactions were as follows:

$$4NaOH + 2SO_2 + O_2 \rightarrow 2Na_2SO_4 + H_2O \\ Ca(OH)_2 + SO_2 \rightarrow CaSO_4 + H_2O \tag{5.9}$$

After the removal of sulfur dioxide (SO_2), the flue gas is treated in the CO_2 absorption unit. The CO2 is removed from the flue gas by means of a chemical solvent (aqueous solution of mono-ethanolamine) with concentration ranging between:

$$2HOC_2H_4NH_2 + H_2O + CO_2 \leftrightarrow \tag{5.10}$$

$$\leftrightarrow HOC_2H_4NHCO_2^- + HOC_2H_4NH_3$$
$$+ 2HOCHNH_2 + HO + CO_2 \leftrightarrow \tag{5.11}$$

$$\leftrightarrow 2HOCHNH^+ + CO_2 + HOC_2H_4NH_2 +$$
$$H_2O + CO_2 \leftrightarrow reactants \tag{5.12}$$

From these studies it was concluded that CO_2 capture and separation from the flue gas by means of an aqueous solution of mono-ethanolamine (MEA) is a feasible system. Additionally, the SO_2 removal (from flue gases) procedure using sodium and calcium hydroxide has been applied and analyzed. The main conclusions of the study are summarized as follows:

> The higher the concentration of CO_2 in the flue gas, the faster it is absorbed by MEA. It was also found that there are several compounds typically present in flue gas, to which MEA absorption is particularly sensitive (e.g., SO_2, H_2S, NO_x, etc.). These gases are known to vary in concentration. To a lesser or greater extent, the abundance of these molecules in the flue gas depend on the composition of the fuel mixture between coal and biomass.

The aim of this investigation was to remove these compounds as much as possible from the flue gas, since they are known to inhibit the ability of the solvent to absorb CO_2. Also, depending on the combustion conditions, NO_x emissions were found to decrease or remain at the same level. It was concluded that a wet scrubber is a viable option for capturing sulfur dioxide.

In aqueous sodium hydroxide solutions, a reduction of SO_2 by 90% and greater was measured. CO_2 concentration in the flue gas has been decreased by 60%, representing an average of all data mapped. However, the low content of sulfur, of oxides of nitrogen, and some particles of ash and dust, which were in the flue gas before the CO_2 absorber, has determined the degradation of MEA.

It was found that large quantities of heat were necessary for the desorption unit to regenerate the MEA solvent.

For instance, power plants with CO_2 capture lose about 10% in efficiency compared with those without CO_2 capture (Alie et al., 2005; Beising, 2007; Cebrucean, 2007). This means that the consumption of fossil fuels will increase dramatically, and the cost of energy production will increase too.

It was estimated that a significant reduction was possible in postcombustion capture costs, from about \$50 down to approximately 20 \$ per ton of CO_2. These results indicated that co-firing biomass with coal is a viable option that can be used to capture CO_2, SO_2, and NO_x. It will also help to achieve clean combustion in addition to capturing CO_2. The process of the co-combustion of a fossil fuel with a renewable CO_2 neutral energy resource (biomass waste) is indeed a very useful application (Ionel et al., 2006, 2007, 2008; Metz et al., 2005; Scheer, 2007; Kohl & Nielson, 1997; Petrova et al., 2006; Thambimuthu et al., 2002). Globally the amount of CO_2 reduction might become more attractive as price decreases, taking into account the CO_2 credits achieved by using this procedure.

Appendix A: Surface Chemistry Essentials

A.1 INTRODUCTION TO SURFACE CHEMISTRY ESSENTIALS

The purpose of this **Appendix** is to provide some basic and essential information about different terms related to surface chemistry as used throughout this book. Some of these terms may be new to some readers, and therefore they are provided in this appendix. It is important to describe the different states (phases) of matter as found in the universe. All the matter that exists around the universe consists of different distinct phases:

gas,

liquid, and

solid phases.

These phases each exhibit specific characteristics.

The gas phase: its molecules fills a container regardless of the shape of the container (Figure. A.1). This is because the gas molecules are free to move (without breaking any bonds) and thus fill the whole volume of any container.

The liquid phase also fills a container corresponding to its volume (contrary to the gas) regardless of shape of the container (Figure A.1).

A solid phase, however, keeps its shape, no matter what container it is placed in (Figure A.1). The molecules in a solid are strongly bonded to each other and cannot move unhindered. The distance between molecules in the solid phase is generally 10% shorter than in the liquid phase.

These considerations are based on macroscopic dimensions. The state of molecules is rather different, as follows. These are called **bulk phases**. In

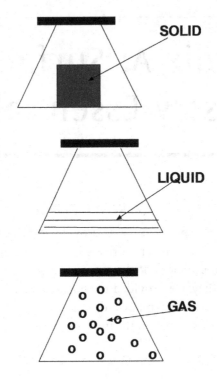

FIGURE A.1   Schematic of solid or liquid or gas in a container.

the bulk phase, each atom/molecule is surrounded by symmetrical neighboring atoms/molecules:

OOOOOOOOOOOO

OOOOOOOOOOOOO

OOOOOOOOOOOO

The molecule (O) is surrounded by the same kinds of neighbors. Thus it interacts with equivalent kind of forces (symmetrical) and is stable (Figure A.2).

The state of the system needs a specific analysis when one considers a molecular snapshot of these phases.

As an example, it is useful to analyze the state of phases in a glass of water. One finds liquid water phase and air (vapor of water) above the liquid phase. The dividing line looks very distinct to naked eye, and it

SOLID

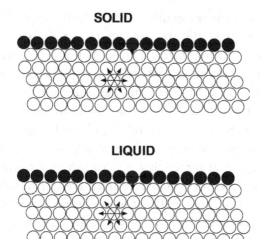

LIQUID

FIGURE A.2 The state of forces acting on a molecule inside a bulk phase and surface (as dark) (e.g., liquid or solid).

becomes evident that at the surface of the liquid water there is a dividing line, the liquid—gas interface.

AIR

WATER (SURFACE) . . . OOOOOOOOOOOO

OOOOOOOOOOOO

OOOOOOOOOOOO   (BULK PHASE)

It is of interest to examine the state of the molecules as (i.e., **O**) situated at the interface. Are these molecules similar (as regards their physical properties) to those molecules (O) as found further deep inside the bulk of liquid phase. The molecules that are situated at the interfaces (e.g., between gas—liquid, gas—solid, liquid—solid,

liquid$_1$—liquid$_2$, solid$_1$—solid$_2$) are *found* to behave differently from those in the bulk phase (Adam, 1930; Harkins, 1952; Davies & Rideal, 1963; Defay et al., 1966; Gaines, 1966; Matijevic, 1976; Chattoraj & Birdi, 1984; Holmberg et al., 2003; Somasundaran, 2006; Birdi, 1997, 2002, 2015). The science related to these interfaces is called **surface chemistry**. It is also well known that the molecules situated near or at the interface (i.e., liquid—gas) are situated differently with respect to each other than the molecules in the bulk phase. Especially, in the case of complex molecules, the orientation in the surface layer will be the major determining factor as regards the surface reactions. The intramolecular forces acting would thus be different in these two cases. Furthermore, it has been pointed out that for a dense fluid, the repulsive forces dominate the fluid structure and are of primary importance. The main effect of the repulsive forces is to provide a uniform background potential in which the molecules move as hard spheres. The attractive forces that stabilize the bulk phase are acting on each molecule are isotropic over a given average time length. Thus, the resultant net force in any direction is absent. The molecules at the interface would be under an asymmetrical force field, which gives rise to the so-called **surface tension** or **interfacial tension** (liquid—liquid; liquid—solid; solid—solid) (Figure A.2). The resultant force on molecules will vary with time because of the movement of the molecules; the molecules at the surface will be pulled downwards into the bulk phase. The presence of this force at the surface molecules will thus give rise to surface tension. This surface force tends to maintain the liquid shape. This becomes obvious when one considers this phenomenon. If one pours a liquid out of a container, it splits up into spherical shapes. The water flowing out of a faucet splits up also into drops. This means that the surface forces are trying to keep the breaking process to a minimum of energy of the system. The minimum energy of the system is achieved by the spherical shape.

The nearer (i.e., at atomic scale ca. 10^{-7} cm) the molecule is to the surface, the greater the magnitude of the force due to **asymmetry.** The region of asymmetry plays a very important role. Accordingly, if the surface area of a liquid is increased, some molecules must move from the interior of the continuous phase to the interface. The surface of a liquid can thus be regarded as the plane of potential energy. An analogous case would be when the solid is crushed and surface area increases per unit gram. Further, molecular phenomena at the surface separating the liquid and the saturated vapor are appreciably more complex than those that occur in the bulk homogeneous fluid phase. In these considerations the *gels* are analyzed as under solid phase.

There are the following systems:

Gas—Liquid

Gas—Solid

Interface is the term used when considering the dividing phase:

Solid—Liquid

$Liquid_1$—$Liquid_2$ (For example: OIL—WATER)

$Solid_1$—$Solid_2$

In other words, the quantity surface tension (γ) may be considered to arise owing to a degree of unsaturation of bonds that occurs when a molecule resides at the surface and not in the bulk. However, the molecules at the surface are easily exchanged with the bulk molecules owing to kinetic movement forces. The term surface tension is used for solid/vapor or liquid/vapor interfaces. The term interfacial tension is more generally used for the interface between two liquids, such as oil—water; two solids, $solid_A$—$solid_B$; or a liquid and solid.

It is, of course, obvious that in a one-component system the fluid is uniform from the bulk phase to the surface, but the orientation of the surface molecules may be different from those molecules in the bulk phase. This means the true description of phases would be thus:

SOLID

INTERFACE SOLID

LIQUID

INTERFACE LIQUID

For instance, one has argued that the orientation of water molecules, H_2O (H—O—H) at the interface most likely is consistent with the oxygen atom pointing at the interface. This would thus lead to a negative dipole, and thus the rain drops would be expected to have a net negative charge (as found from experiments). The question one may ask, then, is how sharply does the density change from that of being fluid to that of gas. Is this transition region a monolayer deep or many layers deep?

Extensive investigations have been reported in the literature over the past decades on this subject (Birdi, 2002, 2015). The Gibbs adsorption theory considers surface of liquids to be of a monolayer thickness. The experiments that analyze the spread monolayers are also based on one molecular layer at the interface. In fact, the latter convincingly confirms the conclusion that the surface is monolayer thick. Further, if one considers a solution made of two (A + B) components, one may find that the composition of A/B in the surface region may or may not be the same as found in the bulk of the solution. The composition of the surface of a solution with two-components or more would require additional comments based on surface chemistry principles.

A.2 SURFACE TENSION OF LIQUIDS

A.2.1 Introduction

Many natural phenomena are related to the characteristics of liquids (for example, water (rain, lakes, rivers, oceans). The surface of oceans is found to play an extensive role in many important natural phenomena. The liquid surface forces are very significant in many of these phenomena. It is important to mention that about 70% of the surface of Earth is covered by water (e.g., oceans, lakes, rivers). The importance of rivers and rain drops on various natural phenomena is very obvious. The effect of rain drops is also made up of dynamic interactions. It is therefore important to give a detailed introduction to the physico-chemical principles of the surface tension of liquids. The most fundamental characteristic of liquid surfaces is that they tend to contract to the smallest surface area in order to achieve the lowest free energy. Whereas gases have no definite shape or volume, completely filling a vessel of any size containing them, liquids have no definite shape but do have a definite volume, which means that a portion of the liquid takes up the shape of that part of a vessel containing it and occupies a definite volume, the free surface being plane except for capillary effects where it is in contact with the vessel. However, if the size of containers is very small, such as in porous materials, then the surface forces (i.e., surface tension of liquids) becomes the dominant parameter. This is observed when one notices rain drops and soap films, in addition to many other systems. The cohesion forces present in liquids and solids and the condensation of vapors to liquid state indicate the presence of much larger intermolecular forces than gravitational forces. Furthermore, the dynamics of molecules at interfaces are important in a variety of areas, such as biochemistry, electrochemistry, and chromatography.

The degree of sharpness of a liquid surface has been the subject of much research in the literature. There is strong evidence that the change in density from liquid to vapor (by a factor of 1000) is exceedingly abrupt, that is, in terms of molecular dimensions. The surface of a liquid was analyzed by light reflectance investigations, as described by Fresnel's law.

Various investigators indeed found that the surface transition involves just one layer of molecules. In other words, when one mentions surfaces and investigations related to this part of a system, one actually mentions just a molecular layer. However, there exists one system which clearly shows the *one molecule thick* layer of surface as being the surface of a liquid: this the monolayer studies of lipids spread on water. The surface thermodynamics of these monolayers is based on the uni-molecular layer at the interface, which thus confirms the thickness of the *surface*. The molecules of a liquid in the bulk phase are in a state of constant unordered motion like those of a gas, but they collide with one another much more frequently owing to the greater number of them in a given volume (as shown here):

GAS PHASE   molecules in gas

    (INTERMEDIATE PHASE)

LIQUID SURFACE   surface molecules

BULK LIQUID PHASE   molecules inside liquid

It is known that a gas molecule occupies 1000 times more volume than a molecule in liquid phase. Further, the *intermediate* phase is only present between the gas phase and the liquid phase. Although one does not often think about how any interface behaves at equilibrium, the liquid surface demands special comment. The surface of a liquid is under constant agitation; there are few things in nature presenting an appearance of more complete repose than a liquid surface at rest. However, according to the kinetic theory, the molecules are subject to much agitation. This is apparent if one considers the number of molecules that must evaporate each second from the surface in order to maintain the vapor pressure. At equilibrium the number of liquid molecules that evaporate into the gas phase is equal to the number of gas molecules that condense at the liquid surface (which will take place in the intermediate phase). The number of molecules hitting the liquid surface is considered to condense irreversibly. From the kinetic theory of gases, this quantity can be estimated as follows:

$$\text{mass} / \text{cm}^2 / \text{second} = \rho_G (k_B T / 2\pi m_m)^{0.5} = 0.0583 \, p_{vap} \left(M_w / T \right) \quad \text{(A.1)}$$

where k_B is the Boltzmann constant (1.3805×10^{-16} erg deg^{-1}), m_m is the mass of molecule, ρ_G is the density of the gas, and M_w is the molecular weight.

For example, in the case of water, at 20°C the vapor pressure of this liquid is 17.5 mm, which gives 0.25g/sec/cm² (from equation A.1). This corresponds to 9×10^{21} molecules of water per second. From consideration of the size of each water molecule one finds that there are ca. 10^{15} molecules, so that it can be concluded that the average life of each molecule in the surface is only about one eight-millionth of a second (1/8 10^{-6} sec). This must be compounded with the movement of the bulk water molecules toward the surface region. It thus becomes evident that there is an extremely violent agitation in the liquid surface. In fact, this turbulence may be considered analogous to the movement of the molecules in the gas phase. One observes this vividly in a cognac glass. The ethanol (in cognac) molecules evaporate and condense on the walls of the container.

In the case of interface between two immiscible liquids due to the presence of interfacial tension, the interface tends to contract. The magnitude of interfacial tension is always lower than the surface tension of the liquid with the higher tension. The liquid—liquid interface has been investigated by specular reflection of X-rays to gain structural information at molecular (Angstrom ($Å = 10^{-8}$ cm = 0.10 nm) resolution (Adamson & Gast, 1997; Birdi, 2002, 2015).

The term **capillarity** (a Latin word capillus: a hair) describes the rise of liquids in fine glass tubes. The rise of fluids in a narrow capillary was related to the difference in pressure across the interface and the surface tension of the fluid:

$$\Delta P = \gamma(\text{curvature}) = \gamma(1/\text{radius of curvature})$$
$$= 2\,\gamma(1/\text{radius of the capillary}) \tag{A.2}$$

This means that when a glass tube of a hair-fine diameter is dipped in water, the liquid meniscus will rise to the very same height. A fluid will rise in the capillary if it wets the surface, while it will decrease in height if it non-wets (like Hg in glass capillary) (Adamson & Gast, 1997; Birdi, 2015). The magnitude of rise is rather large, that is, 3 cm if the bore is of 1 mm for water. This equation also explains what happens when liquid drops are formed at a faucet. Thus any curved liquid surface (Figure A.3) exhibits capillary force. Although it may not be obvious here, but the capillary force can be very dominating in different processes (for example, the properties of a sponge or oil/gas recovery from a reservoir).

In Figure A.3 it is found that the rise of liquid takes place due to the ΔP only, since the liquid meniscus is curved. The curvature induces ΔP across the interface and liquid rises, corresponding to the magnitude of ΔP.

**CAPILLARY PRESSURE
(FLAT & CURVED LIQUID SURFACE)**

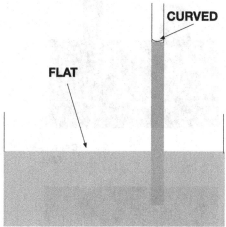

FIGURE A.3   A flat surface ($\Delta P = 0$) and a curved liquid surface (ΔP = capillary pressure).

There is practically no gravitational force present. In porous solid materials (such as sponges, soil, gas/oil/shale reservoir), this capillary force thus becomes the most significant driving force (for example, oil/gas reservoirs, ground water seepage, sponge, fabrics, etc.). The gas adsorption in porous solids exhibits capillary condensation phenomena (Adamson & Gast, 1997; Birdi, 1997, 2002; Myers et al., 2002).

The magnitude of capillary rise is higher in the smaller tubing than in the larger, since the magnitude of ΔP is higher in the former. This explains why it is difficult to recover oil from some reservoirs (such as shale reservoirs, which have very small pores). The same is found in the case of two bubbles or drops (Figure A.5), where the smaller bubble or drop (due to lager ΔP) will coalescent with the larger bubble or drop.

The capillary phenomenon thus means that it will be expected to play an important role in all kinds of systems where liquid (with curved surface) is in contact with materials with pores or holes. In such systems the capillary forces will determine the characteristics of liquid—solid systems. Some of the most important are thus:

- all kinds of fluid flow inside solid matrices (ground water, oil recovery)

- fluid flow inside capillary (oil recovery, ground water flow, blood flow)

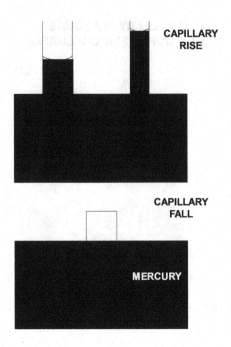

FIGURE A.4 Capillary rise of liquid due to Laplace pressure (ΔP). (In the case of mercury (Hg) there is a fall due to contact angle is greater than 90°).

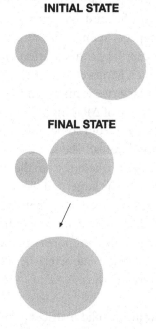

FIGURE A.5 The smaller drop (or bubble) will merge into the larger drop due to the difference in the Laplace pressure.

It was recognized at a very early stage that only the forces from the molecules in the surface layer act on the capillary rise. An impressive example from everyday life is the flow of blood in all living species, which is dependent on capillary forces. The oil recovery technology in reservoirs (shale gas/oil) is similarly dependent on the capillary phenomena. The capillary forces become very dominating in latter systems.

Furthermore, virtually all elements and chemical compounds have a solid, liquid, and vapor phase. A transition from one phase to another phase is accompanied by a change in temperature, pressure, density, or volume. This observation thus also suggests that because of the term ΔP, the chemical potential will be different in systems with flat surfaces.

A typical molecular explanation can be useful to consider in regard to surface molecules. Molecules are small objects that behave as if of definite size and shape in all states of matter (e.g., gas [G], liquid [L], and solid [S]). The volume occupied by a molecule in the gas phase is some 1000 times larger than the volume occupied by a molecule in the liquid phase, as follows:

As shown above, the volume of one mole of a substance (, for example, water in the gas phase (at standard temperature and pressure), V_G (ca. 24,000 cc/mol) is some 1000 times its volume in the liquid phase, V_L (molar volume of water = ca. 18 cc/mol). The distance between molecules, D, will be proportional to $V^{1/3}$ such that the distance in the gas phase, D_G, will be approximately 10 (= $1000^{1/3}$) times larger than in the liquid phase, D_L.

The finite compressibility and the relatively high density, which characterize liquids in general, point to the existence of repulsive and attractive intermolecular forces. The same forces that are known to be present in the gaseous form of a substance may be imagined to play a role also in the liquid form. The mean speed of the molecules in the liquid is the same as that of the molecules in the gas; at the same temperature, the liquid and gas phase differ mainly by the difference in the density between them. The magnitude of surface tension, γ, is determined by the internal forces in the liquid; thus, it will be related to the internal energy or cohesive energy.

It has been explained that the difference between the molecules at the surface and bulk arises from the number of near neighbors (Figure A.2). In other words, that there should be a correlation between the heat of evaporation and surface tension. This was indeed found. The quantities ratio (i.e., surface tension, heat of evaporation) was found to be approximately ½ (as expected from the Figure A.2).

The phenomenon of surface tension can be explained by assuming that the surface behaves like a stretched membrane, with a force of tension acting in the surface at right angles, which tends to pull the liquid surface away from this line in both directions. Surface tension thus has units of force/length = mass distance/time 2 distance = mass/time². This gives surface tension in units as mN/m, dyn/cm, or Joule/m² (mN m/m²).

For example (water): Volume per Mole in Gas or Liquid Phase and Distance Between Molecules in Gas and Liquid Phases

Molar volume of water (at 20°C):

$$V_{gas} = \text{ca. 24,000 ml as gas}$$

$$V_{liquid} = \text{ca. 18 ml as liquid}$$

Ratio:

$$V_{gas} : V_{Liquid} = \text{ca. } 1000 \tag{A.3}$$

Distance (D) between molecules in gas (D_G) or liquid (D_L) phase:

$$\text{Ratio } D_G : D_L = \left(V_G : V_L\right)^{1/3} = \left(1000\right)^{1/3} = 10 \tag{A.4}$$

This shows that the quantity surface tension, γ, as the free energy excess per unit area:

$$\gamma = G_{surface} / \text{area} \tag{A.5}$$

where $G_{surface}$ is the free energy of the two-phase system (phases a and b). The liquid and vapor phases are separated by a surface region (Adamson & Gast, 1997; Birdi, 2002, 2015).

It is also seen that other thermodynamic quantities would be given as:

$$\text{surface energy} = U_{surface} = U / \text{area} \tag{A.6}$$
$$\text{surface entropy} = S_{surface} = S / \text{area} \tag{A.7}$$

and from this one can obtain:

$$\gamma = U_{surface} - S_{surface} \tag{A.8}$$

Hence, the magnitude of surface tension is thus equal to the work needed in forming unit surface area (m^2 or cm^2). This work increases the potential energy or free surface energy, G_s ($J/m^2 = erg/cm^2$), of the system. This can be further explained by different observations one makes in everyday life, where liquid drops contract to attain minimum surfaces. It is well known that the attraction between two portions of a fluid decreases very rapidly with the distance and may be taken as zero when this distance exceeds a limiting value, R_c, the so-called range of molecular action. Analyses have shown that (Rowlinson & Widom, 2003; Chattoraj & Birdi, 1984; Birdi, 2015) surface tension, γ, is a force acting tangentially to the interfacial area, which equals the integral of the difference between the external pressure, p_{ex}, and the tangential pressure, p_t:

$$\gamma = \int \left(p_{ex}\, p_t \right) dz \tag{A.9}$$

the z-axis is normal to the plane interface and goes from the liquid to the gas. The magnitude of work that must be used to remove a unit area of a liquid film of thickness t will be proportional to the tensile strength (latent heat of evaporation) of the liquid thickness. In the case of water, this would give approximately 25,000 atm of pressure (600 cal/gm = ca. 25.2×10^9 erg = 25,000 atm).

A.2.2 Heat of Surface Formation and Heat of Evaporation of Liquids

The thermodynamics of surface tension of liquids require the analyses of heat of surface chemistry formation. As mentioned earlier, energy is required to bring a molecule from the bulk phase to the surface phase of a liquid. In the bulk phase, the number of neighbors (six near-neighbors for hexagonal packing and if considering only two-dimensional packing) will be roughly twice the molecules at the surface (three near neighbors, when discounting the gas phase molecules) (See Figure A.2).

The interaction between the surface molecules and the gas molecules will be negligible, since the distance between molecules in the two phases will be very large. Furthermore, as explained elsewhere, these interaction differences disappear at the critical temperature. It was argued that when a molecule is brought to the surface of a liquid from the bulk phase (where each molecule is symmetrically situated with respect to each other), the work done against the attractive force near the surface will be expected to be related to the work spent when it escapes into the vapor phase. It can be shown that this is just half for the vaporization process (Figure A.6).

MOLECULAR FORCES

MOLECULES AT A SURFACE

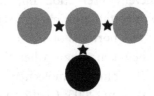

MOLECULES INSIDE BULK PHASE

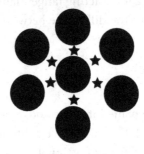

FIGURE A.6   Molecular packing in two-dimensions in bulk and surface (shaded) molecules (schematic).

One finds in the literature a correlation between the latent heat of evaporation, L_{evap}, and γ or the specific cohesion, a^2_{co} ($2\gamma/\rho_L = 2\gamma v_{sp}$), where ρ_L = density of the fluid and v_{sp} is the specific volume. The following correlation was given:

$$L_{evap}\left(V_m\right)^{3/2}/a^2_{co} = 3 \qquad (A.10)$$

However, later analyses showed that this correlation was not very satisfactory for experimental data. From these analyses it was suggested that there are 13,423,656 layers of molecules in 1 cm³ of water.

It is well known that both the heat of vaporization of a liquid, ΔH_{vap}, and the surface tension of the liquid, γ, are dependent on temperature and pressure, and they result from various inter-molecular forces existing within the molecules in the bulk liquid. In order to understand the molecular structure of liquid surfaces, one may consider this system in a somewhat simplified model.

The amount of heat required to convert 1 gm of a pure liquid into saturated vapor at any given temperature is called the latent heat of evaporation or latent heat of vaporization, L_{evap}. It has been suggested that

$$\text{latent heat of evaporation} / 2\gamma = L_{evap} / 2\gamma \qquad (A.11)$$

$$= \text{area occupied by all molecules if they lie in the surface} = A_{mol} \quad (A.12)$$

From this model one can derive the following:

$$\text{diameter } A_{mol} = v_{sp} \qquad (A.13)$$

Thus,

$$\text{diameter} = 2gv_{sp} / L_{evap} \qquad (A.14)$$

For example, the analyses for water is thus:

$$L_{evap} = 600 \text{ g cal } (Kg\ J)$$
$$= 600 \times 42{,}355 \text{ g cm} = 25{,}413{,}000 \text{ g cm}$$
$$v_{H2O} = \text{ca. 1 g/cc} \qquad (A.15)$$
$$\ddagger_{oC} = 88 \text{ dyn/cm} = 0.088 \text{ N/m}$$

From this one finds

$$\text{diameter of water molecule} = 2 \times 0.088 \times 1 / (2{,}541{,}300)$$
$$= 0.7\ 10^{B^8} \text{ cm} = 0.7\ \overset{\circ}{A} = 0.07 \text{ nm} \qquad (A.16)$$

which is of the right order of magnitude.

In a later investigation, a correlation between heat of vaporization, ΔH_{vap}, and the effective radius of the molecule, R_{eff}, and surface tension, γ, was found. These analyses showed that a correlation between enthalpy and surface tension exists that is dependent on the size of the molecule. It thus confirms the molecular model of liquids.

A2.2.3 Surface Tension of Liquid Mixtures

All industrial and natural liquid systems are made up of more than one component, which makes the studies of mixed liquid systems important. Further, the natural oil consists of a variety of alkanes (besides other organic molecules). The analyses of surface tension of liquid mixtures (for example, two or three or more components) has been the subject of studies in many reports. According to one of these models of liquid surfaces, the free energy of the molecule is given as (Adamson & Gast, 1997; Birdi, 1997, 2002, 2015):

$$G_i = k_B T \ln (a_i) \qquad (A.17)$$

where a_i is the absolute activity. This latter term can be expressed as

$$a_i = N_i \, g_i \qquad (A.18)$$

where N_i is the mole fraction (unity for pure liquids) and g_i is derived from the partition function.

The free energy can thus be rewritten as:

$$\begin{aligned} G_i &= g_i \, s_i \\ &= k_B T \ln (a_1/a_1^s) \end{aligned} \qquad (A.19)$$

where s_1 is the surface area per molecule. This is the free energy for bringing the molecule, a_1, from the bulk to the surface, a_1. In a mixture consisting of two components, 1 and 2, one can derive the free energy terms as follows for each species:

$$\gamma s_1 = k_B T \ln \left(N_1 g_1 / N_1^s / g_1^s \right) \qquad (A.20)$$

and

$$\gamma s_2 = k_B T \ln \left(N_2 g_2 / N_2^s / g_2^s \right) \qquad (A.21)$$

where N^s is the mole-fraction in the surface such that

$$N_1^s + N_2^s = 1 \qquad (A.22)$$

As a first approximation one may assume that $s = s_1 = s_2$; that is, the surface area per molecule of each species is approximately the same. This will be reasonable to assume in such cases as mixtures of hexane + heptane, for example. This gives:

$$\gamma_s = k_B T \left(\ln \left(N_1 g_1 g_1^s \right) + \ln \left(N_2 g_2 / g_2^s \right) \right) \qquad (A.23)$$

Or, in combination with equation 22, one can rewrite as follows:

$$\exp(-\gamma_s /k_B T) = N_1 \exp(-\gamma_1 s/k_B T) + N_2^2 \exp(-g_2 s/k_B T) \qquad (A.24)$$

Using the regular solution theory, the relation between activities was given as

$$R T \ln f_1 = -a_1 N_2^2; R T \ln f_2 = -a_1 N_1^2 \qquad (A.25)$$

where f_1 denotes the activity coefficient.

In the case of some mixtures, a simple linear relationship has been observed:

(1) iso-Octane—benzene mixtures: The surface tension changes gradually throughout. This means that the system behaves almost as an ideal;

(2) Water—electrolyte mixtures: The surface tension data of water-NaCl mixtures showed that the magnitude of γ **increases** linearly from ca. 72 to 80 mN/m for 0- to 5-M NaCl solution (d γ/d mol NaCl = 1.6 mN/mol NaCl).

The solutions of water—$NH_4 NO_3$ also showed an increase in γ, with the increase in concentration of NH_4NO_3.

It was found that the increase in γ per mol added NaCl is much larger (1.6 mN/m mol) than that for NH_4NO_3 (1.0 mN/m mol). In general, the magnitude of surface tension of water increases on the addition of electrolytes, with a very few exceptions. This indicates that the magnitude of surface excess term is different for different solutes. In other words, the state of solute molecules at the interface is dependent on the solute.

A.2.3 Solubility of Organic Liquids in Water and Water in Organic Liquids

The process of solubility of one compound into another is of fundamental importance in everyday life: examples are industrial applications (paper, oil, paint, washing) and pollution control (oil spills, waste water control, toxicity, biological processes such as medicine, etc.). Accordingly, many reports are found in the literature that describe this process both on a theoretical basis and by using simple empirical considerations. The molecular picture of the system is very important for the understanding of the mechanisms. As already described here, the formation of a surface or interface requires energy; however, how theoretical analyses can be applied to curvatures of a *molecular-sized cavity* is not satisfactorily developed. It is easy to accept that any solubility process is in fact the procedure where a solute molecule is placed into the solvent where a cavity has to be made. The cavity has both a definite surface area and volume. The energetics of this process is thus a surface phenomenon, even if of molecular dimensions (i.e., nm²). Solubility of one compound, S, in a liquid such as water, W, means that molecules of S leave their neighbor molecules (SSS) and surround themselves with WWW molecules. Thus, the solubility process means formation of a *cavity* in the water bulk phase where a molecule, S, is placed (WWWSWWW). It has been suggested that this cavity formation is a surface free energy process for solubility (Birdi, 2002, 2015).

The solubility of various liquids in water and vice versa is of much interest in different industrial and biological phenomena of everyday importance. In any of these applications, one would encounter instances where a prediction of solubility would be of interest. Furthermore, solubilities of molecules in a fluid are determined by the free energy of solvation. In more complicated processes such as catalysis, the reaction rate is related to the desolvation effects. A correlation between the solubility of a solute gas and the surface tension of the solvent liquid has been described, which was based on the curvature dependence of the surface tension for $C_6 H_6$, $C_6 H_{12}$, and CCl_4. This was based on the model that a solute must be placed in a hole (or cavity) in the solvent. The change in the free energy of the system, ΔG_{sol}, transferring a molecule from the solvent phase to a gas phase is then:

$$\Delta G_{sol} = 4 \pi r^2 \gamma_{aq} e_i \tag{A.26}$$

where e_i is the molecular interaction energy. By applying the Boltzmann distribution law

$$c_{gs} / c_g = \exp(-\Delta G_G / k_B T) \tag{A.27}$$

where c_{gs} is the concentration of gas molecules in the solvent phase and c_g is their concentration in the gas phase. Combining these equations, we obtain:

$$\ln(c_g^s / c_g) = (-4 \pi r^2 \gamma_{aq} / k_B T) + e_i / k_B T \tag{A.28}$$

This model was tested for the solubility data of argon in various solvents, where a plot of log (Oswald coefficient) vs. surface tension was analyzed. In the literature, similar linear correlations were reported for other gas (e.g., He, Ne, Kr, Xe, O_2) solubility data.

The solubility of water in organic solvents does not follow any of these aforementioned models.

For instance, while the free energy of solubility, ΔG_{sol}, for alkanes in water is linearly dependent on the alkyl chain, there exists no such dependence of water solubility in alkanes.

A.2.4 The Hydrophobic Effect

All natural processes are in general dependent on the physicochemical properties of water (especially when considering that over 70% of the Earth is covered by water).

Amphiphile molecules, such as long chain alcohols or acids, detergents, lipids, or proteins, exhibit polar-apolar characteristics, and the dual behavior is given this designation. The solubility characteristics in water are determined by the alkyl or apolar part of these amphiphiles, which arise from hydrophobic effect. Hydrophobicity plays an important role in a wide variety of phenomena, such as solubility in water of organic molecules, oil—water partition equilibrium, detergents, washing and all other cleaning processes, biological activity, drug delivery, and chromatography techniques. Almost all drugs are designed with a particular hydrophobicity as determined by the partitioning of the drug in the aqueous phase and the cell lipid membrane.

The ability to predict the effects of even simple structural modifications on the aqueous solubility of an organic molecule could be of great value in the development of new molecules in various fields, for example, medical or industrial. There exist theoretical procedures to predict solubilities of nonpolar molecules in nonpolar solvents and for salts or other highly polar solutes in polar solvents, such as water or similar substances. However, the prediction of solubility of a nonpolar solute in water has been found to require some different molecular considerations (Birdi, 1982).

Furthermore, the central problems of living matter comprise the following factors: recognition of molecules leading to attraction or repulsion, fluctuations in the force of association and in the conformation leading to active or inactive states, the influence of electromagnetic or gravitational fields and solvents including ions, and electron or proton scavengers. In the case of life processes on Earth, one is mainly interested in solubility in aqueous media.

The unusual thermodynamic properties of nonpolar solutes in aqueous phase were analyzed, by assuming that water molecules exhibit a special ordering around the solute. This water-ordered structure was called the *iceberg structure*. The solubility of semipolar and nonpolar solutes in water has been related to the term molecular surface area of the solute and some interfacial tension term.

The solubility, X_{solute}, in water was derived as:

$$R\,T\,(\ln X_{solute}) = -(\text{surface area of solute})\,(\gamma_{sol}) \quad (A.29)$$

where surface tension, γ_{sol}, is some *micro—interfacial tension* term at the solute-water (solvent) interface.

The quantity surface area of a molecule is the cavity dimension of the solute when placed in the water medium.

The data of solubility, total surface area (TSA), and hydrocarbon surface area (HYSA) have been analyzed for some typical alkanes and alcohols. The relationship between different surface areas of contact between the solute solubility (sol) and water were derived as:

$$\ln (sol) = -0.043\,TSA + 11.78/(RT)\Delta G_{o,sol}$$
$$= -RT \ln (sol) = 25.5\,TSA + 11.78 \quad (A.30)$$

where *sol* is the molar solubility and TSA is in Å^2.

The quantity 0.043 (RT = 25.5) is some micro-surface tension. It is also important to mention that at the molecular level there cannot exist any surface property that can be uniform in magnitude in all directions. Hence, the micro-surface tension will be some average value.

In the case of alcohols, assuming a constant contribution from the hydroxyl group, the hydrocarbon surface area (HYSA) = TSA—hydroxyl group surface area:

$$\ln (sol) = -0.0396\,HYSA + 8.94 \quad (A.31)$$

However, one can also derive a relationship that includes both HYSA and OHSA (hydroxyl group surface area):

$$\ln (sol) = -0.043\,HYSA\,(0.06\,OHSA + 12.41 \quad (A.32)$$

The relations described above did not give correlations to the measured data that were satisfactory (ca. 0.4 to 0.978). The following relationship was derived based on the solubility data of both alkanes and alcohols, which gave correlations on the order of 0.99:

$$\ln{(sol)} = 0.043 \text{ HYSA} + 8.003 \text{ IOH} - 0.0586 \text{ OHSA} + 4.42 \quad \text{(A.33)}$$

where the IOH term equals 1 (or the number of hydroxyl groups) if the compound is an alcohol and zero if the hydroxyl group is not present.

The term HYSA thus can be assumed to represent the quantity that relates to the effect of the hydrocarbon part on the solubility. The effect is negative, and the magnitude of t is 17.7 erg/cm^2.

The magnitude of OHSA is found to be 59.2 Å^2. As an example, the surface areas of each carbon atom and the hydroxyl group in the molecule 1-nonanol were estimated.

$$CH_3CH_2CH_2CH_2CH_2CH_2CH_2CH_2OH$$

84.9/31.8/31.8/31.8/31.8/31.8/31.8/31.8/31.8/OH

It is seen that the surface area of the terminal methyl group (84.9 Å^2) is approximately three times larger than the methylene groups (31.82 Å^2, or 31.82 10^{-20} m^2). Computer simulation techniques have been applied to such solution systems.

A.3 INTERFACIAL TENSION OF LIQUIDS (LIQUID$_1$— LIQUID$_2$, OIL—WATER)

A.3.1 Introduction

The interfacial forces present between two phases, such as immiscible liquids, are of much importance from a theoretical standpoint, as well as in regard to practical systems. The liquid$_1$—liquid$_2$ interface is an important one as regards such phenomena as chemical problems, extraction kinetics, phase transfer, emulsions (oil-water), fog, and surfactant solutions. In the case of primary oil production, one has to take into consideration the surface tension of oil. However, during a secondary or tertiary recovery, the interfacial tension between the water phase and oil phase becomes an important parameter. For example, the *bypass* and other phenomena such as snap-off are related to the interfacial phenomena (Birdi, 2015).

Interfacial tension (IFT) between two liquids is less than the surface tension of the liquid with the higher surface tension, because the molecules of each liquid attract each other across the interface, thus diminishing the inward pull exerted by that liquid on its own molecules at the surface.

The precise relation between the surface tensions of the two liquids separately against theory vapor and the interfacial tension between the two liquids depends on the chemical constitution and orientation of the molecules at the surfaces. In many cases, a rule proposed by Antonows holds true with considerable success (Birdi, 2002).

A.3.2 Liquid—Liquid Systems: Work of Adhesion

The surface tension is the force that is present between two different phases (Adamson & Gast, 1997; Birdi, 2002, 2015). The free energy of interaction between dissimilar phases is the work of adhesion, WA (energy per unit area):

$$W_A = W_{AD} + W_{AH} \tag{A.34}$$

where W_A is expressed as the sum of different intermolecular forces, for example,

    a. London dispersion forces, D;
    b. Hydrogen bonds, H;
    c. Dipole-dipole interactions, DD;
    d. Dipole-induced interactions, DI;
    e. Π bonds, Π;
    f. Donor-acceptor bonds, DA;
    g. Electrostatic interactions, EL.

It is also easily seen that the W_{AD} term will always be present in all systems (i.e., liquids and solids), while the other contributions will be present to a varying degree as determined by the magnitude and nature of the dipole associated with the molecules. In order to simplify the terms given by the above equation, one procedure has been to compile all the intermolecular forces arising from the dipolar nature of W_{AP}:

$$W_A = W_{AD} + W_{AP} \tag{A.35}$$

where

$$W_{AP} = W_{AH} + W_{ADD} + W_{AID} \tag{A.36}$$

The calculated value of surface tension of n-octane was analyzed from these parameters.

The calculated value for γ of octane = 19.0 mN/m, while the measured value is 21.5 mN/m, at 20°C (i.e., $\gamma_{octane} = \gamma_{LD}$). The real outcome of this example is that such theoretical analyses do indeed predict the surface dispersion forces, γ_{LD}, as measured experimentally, to a good accuracy. In a further analysis, the Hamaker constant, A_i, for liquid alkanes is found to be related to γ_{LD} as:

$$A_i = 3 \times 10^{-14} (\gamma_{LD})^{11/12} \tag{A.37}$$

This was further expanded to include components at an interface between phases I and II:

$$A_{I,II} = 3 \times 10^{-14} / e_2 (\gamma_I{}^D - \gamma_{II}{}^D)^{11/6} \tag{A.38}$$

where e_2 is the dielectric constant of phase 2; however, in some cases, forces other than dispersion forces would also be present. The manifestation of intermolecular forces is a direct measure of any interface property and requires a general picture of the different forces responsible for bond formation, as discussed in the following.

a. Ionic bonds: The force of attraction between two ions is given as:

$$F_{ion} = (g^+ g^-) / r^2 \tag{A.39}$$

and the energy, U_{ion}, between two ions is related to r_{ion} by the equation

$$U_{ion} = (g^+ g^-) / r_{ion} \tag{A.40}$$

where two charges (g^+, g^-) are situated at a distance of r_{ion}.

b. Hydrogen bonds: Based on molecular structure, those conditions under which hydrogen bonds might be formed are (a) presence of a highly electronegative atom, such as O, Cl, F, and N, or a strongly electronegative group such as -CCl₃ or -CN, with a hydrogen atom attached; (b) in the case of water, the electrons in two unshared sp³ orbitals are able to form hydrogen bonds; (c) two molecules such as $CHCl_3$ and acetone (CH_3COCH_3) may form hydrogen bonds when mixed with each other, which is of much importance in interfacial phenomena.

c. Weak-electron sharing bonding: In magnitude this is of the same value as the hydrogen bond. It is also the Lewis acid-Lewis base bond (comparable to Brønsted acids and bases). Such forces might contribute appreciably to cohesiveness at interfaces; a typical example is the weak association of iodine (I_2) with benzene or any polyaromatic compound.

The interaction is the donation of the electrons of I_2 to the electron-deficient aromatic molecules (π-electrons).

d. Dipole-induced dipole forces: In a symmetrical molecule, such as CCl_4 or N_2, there is no dipole (ma = 0) through the overlapping of electron clouds from another molecule with dipole, m_b, with which it can interact with induction. It will thus be clear that various kinds of interactions would have to be taken into consideration whenever we discuss interfacial tensions of liquid—liquid or liquid—solid systems (Adamson & Gast, 1997; Kwok et al., 1994).

A.3.3 Interfacial Tension Theories of Liquid—Liquid Systems

As shown above, various types of molecules exhibit different intermolecular forces, and their different force and potential-energy functions can be estimated (Birdi, 1997, 2002). If the potential-energy function were known for all the atoms or molecules in a system, as well as the spatial distribution of all atoms, it could in principle then be possible to add up all the forces acting across an interface.

Further, this would allow one to estimate the adhesion or wetting character of interfaces. Because of certain limitations in the force field and potential-energy functions this is not quite so easily attained in practice. Further, the microscopic structure at a molecular level is not currently known.

For example, to calculate the magnitude of surface tension of a liquid, one needs knowledge of the radial pair-distribution function. However, for the complex molecule, this would be highly difficult to measure, although data for simple liquids such as argon have been found to give the desired result. The intermolecular force in saturated alkanes arise only from London dispersion forces. Now, at the interface, the hydrocarbon molecules are subjected to forces from the bulk molecule, equal to γ. Also, the hydrocarbon molecules are under the influence of London forces due to molecules in the oil phase. It has been suggested that the most plausible model is the geometric means of the force due to the dispersion attraction, which should predict the magnitude of the interaction between any dissimilar phases.

As described earlier, the molecular interactions arise from different kinds of forces, which means that the measured surface tension, γ, arises from a sum of dispersion, γ_D, and other polar forces, γ_P (Chapter 3):

$$\gamma = \gamma_D + \gamma_P \tag{A.41}$$

Here, γ_D denotes the surface tensional force solely determined by the dispersion interactions, and

γ_P arises from the different kinds of polar interactions (Equation 66). The interfacial tension between hydrocarbon (HC) and water (W) can be written as:

$$\gamma_{HC,W} = \gamma_{HC} + \gamma_W - 2 \left(\gamma_{HC} \gamma_{W,D}\right)^{1/2} \tag{A.42}$$

where subscripts HC and W denote the hydrocarbon and water phases, respectively. Considering the solubility parameter analysis of mixed-liquid systems, it is found that the geometric mean of the attraction forces gives the most useful prediction values of interfacial tension. Analogous to that analysis in the bulk phase, the geometric mean should also be preferred for the estimation of intermolecular forces at interfaces. The geometric mean term must be multiplied by a factor of two since the interface experiences this amount of force by each phase. However, the relation in Equation A.43 was alternatively proposed by Antonow (Birdi, 1997, 2002):

$$\gamma_{12} = \gamma_1 + \gamma_2 - 2 \left(\gamma_1 \gamma_2\right)^{1/2} = \left((\gamma_1)^{1/2} - (\gamma_2)^{1/2}\right)^2 \tag{A.43}$$

This relation is found to be only an approximate value for such systems as fluorocarbon—or hydrocarbon—water interfaces, while not applicable to polar organic liquid—water interfaces.

In order to analyze these systems, a modified theory was proposed (Chapter 3). The expression for interfacial tension was given as (Adamson & Gast, 1997; Birdi, 1997):

$$\gamma_{12} = \gamma_1 + \gamma_2 - 2 \, \Phi \left(\gamma_1 \gamma_2\right)^{1/2} \tag{A.44}$$

where the value of Φ varied between 0.5 and 0.15. Φ is a correction term for the disparity between molar volumes of v_1 and v_2. This theory was extensively analyzed in the literature, and satisfactory agreement was found with experimental data.

A.3.4 Analysis of the Magnitude of the Dispersion Forces in Water (γ_D)

Water is known to play a very important role in a variety of systems encountered in everyday life, and its physicochemical properties are of much interest. Therefore, the magnitude of water γ_D has been the subject of much investigation and analysis. By using Equation A.42 and the measured data of interfacial tension for alkanes-water, the magnitude of γ_D has generally been accepted to be 21.8 mN/m (at 25°C).

In order to obtain any thermodynamic information of such systems it is useful to consider the effect of temperature on IFT. The alkane-water IFT data has been analyzed.

These data show that IFT is lower for C_6H_{14} (50.7 mN/m) than for the other higher-chain-length alkanes. The slopes (interfacial entropy: $-d\gamma$ / dT) are all almost the same, ca. 0.09 mN/m per CH_2 group. This means that water dominates the temperature effect, or that the surface entropy of IFT is determined predominantly by the water molecules. Further, as described earlier, the variation of surface tension of alkanes varies with chain length. This characteristic is not present in IFT data; however, it is worth noticing that the slopes in IFT data are lower than that of both pure alkanes and water.

A.4 LIQUID—SOLID SYSTEMS (CONTACT ANGLE— WETTING—ADHESION) UNDER DYNAMIC CONDITIONS

The state of liquid in contact with a solid surface is of much importance in many everyday phenomena (detergency, adhesion, friction, wetting, flotation, suspensions, solid emulsions, erosion, printing, pharmaceutical products, oil/gas reservoirs, etc.). If one considers two systems, such as a drop of liquid (water) placed on different solid surfaces (glass, Teflon), one observes the following. The contact angle, θ, (Figure A.7), as defined by the balance between surface forces (surface tensions) between the respective phases, solid (γ_S), liquid (γ_{liquid}), and LS (liquid—solid) (γ_{SL}) (Chapter 3) (Birdi et al., 1989; Birdi & Vu, 1993; Schwartz, 1999):

$$\gamma_S = \gamma_{SL} + \gamma_{liquid} \, Cos(\theta) \tag{A.45}$$

In the case of water—glass and water—Teflon, one finds that the magnitude of θ is 30° and 105°, respectively. Since the liquid is the same, then the difference in contact angle arises from the different solid surface tensions. From this one can therefore conclude that the surface tension of a solid is

**BALANCE OF SURFACE
TENSIONS AT A CONTACT ANGLE (CA)
(LIQUID - SOLID SYSTEM)**

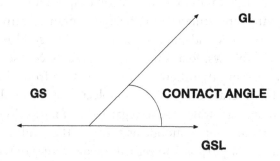

FIGURE A.7   Contact angle (θ = CA) at the liquid—solid interface.

an important surface parameter (Chapter 3). Further, the relation between Young's equilibrium contact angle and the **hysteresis** on rough paraffin wax surfaces was investigated.

Advancing and receding contact angles (contact angle hysteresis) of four organic liquids and water were measured on a variety of polymer surfaces and silicon wafers using an inclinable plane. The magnitude of contact angles varied widely from liquid to liquid and from surface to surface. Surface roughness was relatively unimportant. Instead, the contact angles seemed to be more closely tied to the chemical nature of the surfaces. In general, contact angles increased with the liquid surface tension and decreased with the surface tension of the solid. Several definitions were used to calculate contact angle hysteresis from the experimental data. Although hysteresis is usually considered an extensive property, it was found that on a given surface a wide range of liquids gave a unique value of reduced hysteresis. Apparently, reduced hysteresis represents an intrinsic parameter describing liquid-solid interactions.

In some studies, the **dynamic systems of liquid drops,** when placed on a smooth solid surface, have been investigated. The system liquid drop—solid is a very important system in everyday life. For example, this is significant in rain drops placed on tree leaves or other surfaces. It is also significant in all kinds of systems where a spray of fluid is involved, such as in sprays or combustion engines. Further, in many diagnostics devices (such as blood analyses), enzymes are applied in very small amounts (microliters) from solutions as a drop on an electrode. This means that after evaporation of the fluid the remaining enzyme must be well described for the diagnostic instrument to function reliably. The dynamics of a liquid drop evaporation rate is of much interest in many phenomena (combustion engines, rain drops and environment, aerosols and pollution). The liquid—solid interface can be considered as follows. Real solid surfaces are, of course, made up of molecules not essentially different in their nature from the molecules of the fluid. The interaction between a molecule of the fluid and a molecule of the boundary wall can be regarded as follows. The molecules in the solid state are not as mobile as those of the fluid. It is therefore permissible for most purposes to regard the molecules in the solid state as stationary. However, complexity arises in those liquid—solid systems where a layer of fluid might be adsorbed on the solid surface, such as in the case of water—glass.

Systematic studies have been reported in literature on the various modes of liquid drop evaporation when placed on smooth solid surfaces (Birdi, 2002, 2015). In these studies the rate of the mass and contact diameter of water and n-octane drops placed on glass and Teflon surfaces were investigated. It was found that the evaporation occurred with a constant spherical cap geometry of the liquid drop. The experimental data supporting this were obtained by direct measurement of the variation of the mass of droplets with time and by the observation of contact angles. A model based on the diffusion of vapor across the boundary of a spherical drop was considered to explain the data. Further studies were reported, where the contact angle of the system was $\theta < 90$. In these systems, the evaporation rates were found to be linear and the contact radius constant. In the latter case, with $\theta > 90°$, the evaporation rate was non-linear, the contact radius decreased, and the contact angle remained constant.

As a model system, one may consider the evaporation rates of fluid drops placed on polymer surfaces in still air. The mass and evaporating liquid (methyl acetoacetate) drops on polytetrafluoroethylene (Teflon) surface in still air have been reported. These studies suggested two pure modes of evaporation: at constant contact angle with diminishing contact area and at constant contact area with diminishing contact angle. In this mixed mode the drop shape would vary, resulting in an increase in the contact angle with a decrease in the contact circle diameter, or, sometimes a decrease in both quantities.

The data for the state of a liquid drop placed on a smooth solid surface can be described in terms of the

radius,

height of the drop,

weight,

and the contact angle, θ.

The liquid drop when placed on a smooth solid can have a spherical cap shape, which will be the case in all cases where the drop volume is approximately 10 µL or less. In the case of much larger liquid drops, ellipsoidal shapes may be present, and different geometrical analysis will have to be implemented. In the case of rough surfaces, one may have much difficulty in explaining the dynamic results. However, one may also expect that there will be instances where it is non-spherical. This parameter will need be determined before any analyses can be carried. In the case of a liquid drop that is sufficiently small, surface tension dominates over gravity. The liquid drop can then be assumed to form a spherical cap shape. A spherical cap shape can be characterized by four different parameters, the drop height (h_d), the contact radius (r_b), the radius of the sphere forming the spherical cap (R_s), and the contact angle (θ). By geometry the relationships between the two radii, the contact angle, and the volume of the spherical cap (V_c) are given as[125]

$$r_b = R_s \operatorname{Sin}(\theta) \tag{A.46}$$

and

$$R_s = \left((3\,V_c)/(p_b) \right)^{1/3} \tag{A.47}$$

where

$$b = (1 - \cos\theta)^2 (2 + \cos\theta) = 2 - 3\cos\theta + \cos 3\theta \tag{A.48}$$

The height of the spherical cap above the supporting solid surface is related to the two radii and the contact angle, θ, by:

$$h = R_s (1 - \cos\theta) \tag{A.49}$$

and

$$h = r_b \tan (\theta/2) \tag{A.50}$$

A spherical cap-shaped drop can be characterized by using any two of these four parameters. When the horizontal solid surface is taken into account, the rate of volume decrease by time is given as:

$$-(d\, V_c/\, dt) = (4\, p\, R_s D)/(r_L)(c_S - c_4)\, f(\theta) \tag{A.51}$$

where t is the time (s), D is the diffusion coefficient (cm²/sec), c_S is the concentration of vapor at the sphere surface (at R_s distance) (g/cm³), c_4 is the concentration of the vapor at infinite distance (R_s distance) (g/cm³), r_L is the density of the drop substance (g/cm³), and $f(\theta)$ is a function of the contact angle of the spherical cap. In literature one finds a few solutions to this relationship with the analogy between the diffusive flux and electrostatic potential, the exact solution has been derived.

The approximate solution for $f(\theta)$ was given as (Birdi et al., 1993)

$$F(\theta) = (1 - Cos(\theta))\, /\, 2 \tag{A.52}$$

while other investigators gave the following relationships

$$f(\theta) = (Cos\,(\theta))/(2\, \ln\,(1 - Cos(\theta))) \tag{A.53}$$

The data showed that only in some special cases the magnitude of θ remains constant under evaporation, such was in a water—teflon system (Figure A.8).

In this latter case, where contact angle remains constant during evaporation, the following relation can be written:

$$V_c{}^{2/3} = V_{ct}{}^{2/3} - 2/3\, K\, f(\theta)\, t \tag{A.54}$$

The liquid film that remains after most of the liquid has evaporated has been investigated (Birdi et al., 1993). It was shown that one could estimate the degree of porosity of solid surfaces from these data. This method just shows that one can determine the porosity of a solid in a very simple manner as compared to other methods. These data show that one can

**EVAPORATION PROFILES
OF LIQUID (WATER) DROP ON SOLID**

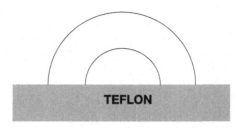

TEFLON

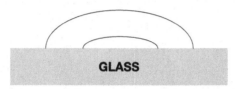

GLASS

FIGURE A.8   Profiles of liquid drops under evaporation (snap shots at two time intervals): water on Teflon; water on glass.

determine the porosity of solids, without the use of mercury porosimeter or other method. The latter studies are much more accurate, since they were based on measurements of change of weight of drop vs. time under evaporation.

Appendix B

The purpose of this appendix is to add some extra information, with regard to the gas adsorption and capture aspects of CCS (Chapter2; Chapter 5). This information is also intended to be useful for further research by the reader. The additional data included here will also explain some methods described in the above text. The gas—solid interaction is described by adding some data from literature. These are useful in selecting the most effective solid (adsorbent) or CCS technology.

Gas Adsorption on Solid Surfaces Essentials:

Solid Surfaces

The process of adsorption of a gas molecule on a solid surface has been investigated for about a century (Langmuir, 1918; Bakker, 1815; Chattoraj & Birdi, 1984; Adamson & Gast, 1997; Myers, 1989; Birdi, 2002, 2016; Keller et al., 1992; Somasundaran, 2006; Tovbin, 2017). Molecules in the gas are moving large distances (ten times) than the molecules in the liquid or solid state. Under suitable conditions, gas molecules interact with a solid surface in different ways, as regards the surface forces. When a gas molecule comes close to a solid surface or a liquid surface, the following may take place (Chapter 2):

The gas molecule may bounce back into the gas phase (absence of adsorption);

The molecule may adsorb at the surface (under the field of force of the solid surface atoms);

The gas may also exhibit preference for adsorption on a specific solid site.

Adsorption is marked by an increase in density of a fluid near the surface, for our purposes, of a solid. In the case of gas adsorption, this happens when molecules of the gas approach the surface and undergo an interaction with it, temporarily departing from the gas phase. Molecules in this new condensed phase formed at the surface remain for a period of time and then return to the gas phase. The duration of this stay depends on the nature of the adsorbing surface and the adsorptive gas, the number of gas molecules that strike the surface and their kinetic energy (or collectively, their temperature), and other factors (such as capillary forces, surface heterogeneities, etc.). Adsorption is by nature a surface phenomenon, governed by the unique properties of bulk materials that exist only at the surface.

A gas molecule/atom can interact with a solid (surface) with varying types of interaction forces or mechanisms. These are mentioned as

chemical adsorption (chemisorption)

weak physical adsorption (physisorption)

penetration into pores (porous solids) (absorption).

Adsorption of a gas molecule (from gas phase) thus gives rise to a lowering in entropy on adsorption (Adamson & Gast, 1997; Keller et al., 1992; Bolis et al., 1999, 1997; Birdi, 1997, 2002). The thermodynamics of physical adsorption of gases in porous solids has been investigated. The measurement of the amount of gas adsorbed by a solid is carried out by volumetric or gravimetric techniques based upon the concept of the (Gibbs) surface excess model (Chattoraj & Birdi, 1984; Myers, 1989).

The subject of gas adsorption on solids has been investigated by the principles of surface thermodynamics (Gibbs adsorption theory) (Chattoraj & Birdi, 1984). Basically, surface thermodynamics analyses provide quantitative relationships between phenomena such as the amount adsorbed and the heat and entropy of gas adsorption. However, in the case of adsorption in porous solids, the surface tension (of the solid) (Chapter 3) and surface area are not easily available by any direct method. However, in the case of microporous solids the gas adsorption phenomena will be different. The gas molecules that are adsorbed inside the pores (of varying size) will interact differently than those adsorbed in large pores.

SOLID MATERIALS USED FOR GAS ADSORPTION: DIFFERENT SOLID MATERIALS USED FOR GAS ADSORPTION

The carbon (CO_2 gas) capture process is based on using different kinds of solids. The characteristics of the solid have to be investigated, in order to find the most appropriate for a given process. The different solids used for gas adsorption are described here.

Porous Solid Materials

The surface properties of a solid are dependent on various factors. The most important arises from the size of solid particles. Finely divided solids possess not only a geometrical surface, as defined by the different planes exposed by the solid but also an internal surface due to the primary particles' aggregation. This leads to pores of different size according to both the nature of the solid and origin of the surface. Experiments show that these pores may be circular, square, or other shape. The porous solids

/S/ /S/ /S/ /S/ /S/ /S/ /S

The size of pores is designated as the average value of the width, w (Sing et al., 2005). The width, w, gives either the diameter of a cylindrical pore, or the distance between the sides of a slit-shaped pore.

The smallest pores, with the range of width $w < 20$ Å (2 nm $= 2 \times 10^{-9}$ m) are called *micropores*.

The *mesopores* are in the range of a width in the 20 Å $\leq w \leq 500$ Å (10^{-8}cm) (2 and 50 nm).

The largest pores in the range of width $w > 500$ Å (50 nm) are called *macropores* (Cambell, 1988; Birdi, 2017). The shapes of pores will vary in geometric size and shape (e.g., circular, square, triangular, etc.). The capillary forces in these pores will thus depend both on the diameter but also on the shape.

In general, most solid adsorbents exhibit (for example, like charcoal and silico-alumina) irregular pores with widely variable diameters in a normal shape. Conversely, other materials such as zeolites and clay minerals are entirely micro- or meso-porous, respectively. In other words, the porosity of these materials is due to the primary particles' aggregation, but this is an intrinsic structural property of the solid material (Rabo, 1976; Birdi, 2017).

Different Solid Adsorbents Used for CO_2 (gas) Capture (adsorption)

The literature is very extensive as regards the different kinds of solids used for gas adsorption. Therefore, it is useful to mention a few here for the sake of description of the mechanisms of the process. Obviously, there is a specific requirement for a suitable absorbent of CO_2 (Chapter 5). This requires that the adsorption—desorption characteristics of the gas should be acceptable for the process. For instance, the adsorbed CO_2 (under high pressure) could be desorbed on change of pressure (reduction of pressure). Or stripping with a suitable gas, such as stream with an inert gas (or some similar process). Further, from surface chemistry principles, the adsorbent should exhibit large surface area per weight. The physical adsorption process has many advantages over other methods. It is low energy system. The rate of the adsorption—desorption step is comparatively short.

ZEOLITES USED FOR GAS ADSORPTION: Gas adsorption on solids, such as **zeolites,** has been investigated. Zeolites are available either as natural state or as synthetic crystalline alumino-silicates, the structure of which is based upon a three-dimensional polymeric framework, with nanosized cages and channels (Thomas & Thomas, 1997; Van Santen et al., 1999; Rabo, 1976). The basic building block of such materials, of general formula: $M^{n+} (AlO) (SiO)^{x-} \cdot zHO$, is the [TO] unit with T = Si, Al.

This unit is a tetrahedron centered (as determined from X-ray analyses) on one T atom bound to four O atoms located at the corners; each O atom is in turn shared between two T atoms. These tetrahedral units join each other through T–O–T linkages in a variety of open-structure frameworks characterized by (interconnected) channels and voids that are occupied by cations and water molecules (Lowell et al., 2006; VanSanten et al., 1999; Rabo, 1976; Yang, 2003).

The presence of charge-balancing (extra-framework) cations is required in order to compensate the negative charge of the tetrahedral $[AlO_4]^-$ units in which Al is in isomorphous substitution of Si atoms. The density of charge-balancing cations depends upon the Si:Al ratio (i.e., range from 1 to ∞).

The most significant adsorption characteristics of zeolites arises from the presence of the nanosized cages and channels within the crystalline structure. Further, the shape of the channels (which gives rise to selectivity properties) makes these useful solid materials for different applications: for example, catalysis and in-gas separation processes. The most characteristic property of any porous material is its large surface area (i.e., area/gram of solid: over 3000 m^2/gm), which maximizes the extension of the interface region.

The presence of charge-balancing (extra-framework) cations is required in order to compensate the negative charge of the tetrahedral $[AlO_4]^-$ units in which Al is in isomorphous substitution of Si atoms. The density of charge-balancing cations depends upon the Si:Al ratio, which varies from

1 to ∞ range (a transition from Si → ∞ Al). However, the presence of nano-sized cages and channels within the crystalline structure of zeolites gives to these materials unique molecular size and shape selectivity properties, of great interest in catalysis and gas separation processes.

The most characteristic property of porous materials is their high surface areas (i.e., area/gram of solid), which maximize the extension of the interface region. Furthermore, in both defective and perfect all-silica zeolites the process was entirely reversible upon evacuation of the gas phase. In this latter case NH_3 interacted only *via* hydrogen bond with Si–OH nests. In the defect-free MFI—silicate, which exposes only unreactive siloxane bridges, the interaction was specific in that it was governed by dispersion forces due to the nanopore walls (*confinement effect*).

The process of gas adsorption is both dependent on the characteristics of the gas and the solid (adsorbents). The surface characteristics of the solid need to be suitable for the gas adsorption process.

Carbon is found in many different forms. For example, carbon black (CB) is one form of carbon that is produced by the incomplete combustion of heavy petroleum products such as FCC (fluid catalytic cracking) tar, coal tar, and ethylene cracking tar, and vegetable oil. Carbon black is a form of amorphous carbon that has a high surface-area-to-volume ratio. However, despite that fact, carbon black, due to specific conductivity and mechanical properties, is not being used as a sensing material in gas sensors. Only activated carbon (also called activated charcoal, activated coal, or carbon activates) is used in gas sensors where CB can be used as a filter. Activated carbon is a form of carbon that has been processed to make it extremely porous and thus to have a very large surface area available for either adsorption or chemical reactions. Due to its high degree of microporosity, just 1 gm of activated carbon has a surface area in excess of 2000 m^2.

Another possibility for carbon black to be integrated in gas sensors is connected with using composites, where another material provides the gas-sensing properties while carbon black plays the part of filler, characterized by high conductivity and high dispersion. The most significant carbon black surface properties useful for composites design are dispersion, stability of the carbon black structure or network, consistent particle size, specific resistance, structure, and purity. Carbon black is used mainly in polymer-based composites. The carbon black endows electrical conductivity to the films, whereas the different organic polymers such as poly(vinyl acetate) (PVAc), polyethylene (PE), poly(ethylene-*co*-vinyl acetate) (PEVA), and poly(4-vinylphenol) (PVP) are sources of chemical diversity between elements in the sensor array. In addition, polymers function as the insulating phase of the carbon black composites. The concentration of CB in composites is varied within the range 2–40 wt%. The conductivity of these materials and their response to compression or expansion can be explained using percolation theory (McLachlan et al., 1990). The compression of a composite prepared by mixing conducting and insulating particles leads to increased conductivity, and, conversely, expansion leads to decreased conductivity.

Micropores, where most adsorption takes place, are in the form of two-dimensional spaces between two graphite-like walls: two-dimensional crystallite planes composed of carbon atoms. The distance between the two neighboring planes of graphite is 3.76 Å **(3.76×10^{-8} cm)** (0.376 nm), but activated carbons have a rather disordered crystallite structure (turbostatic structure). Most activated carbons contain some oxygen complexes that arise from either source materials or from chemical adsorption of air (oxidation) during the activation stage or during storage after activation. Especially, oxides on the surface add a polar nature to activated carbons.

Furthermore, activated carbon also contains to some extent ashes derived from starting materials. The amount of ash varies from 1% to 12%. Ashes are known to consist of silica, alumina, iron, alkaline, and alkaline Earth metals. These ashes exhibit some specific characteristics:

a. an increase in hydrophilicity of activated carbon;
b. catalytic effects of alkaline, alkaline Earth, and some other metals such as iron during activation.

In general one finds that activated carbons in commercial use are present in two forms: powder form and granular or pelletized form.

Decolorization in different refinery processes, removal of organic substances, odor and trace pollutants removal in drinking water treatment, and wastewater treatment are the main applications of liquid phase adsorption.

Carbon molecular sieves: The size of the micropore of the activated carbon is determined during pyrolyzing and activation treatments. Hence, small and defined micropores that have molecular sieving effects can be prepared by using proper starting materials. The main applications of activated carbon with molecular sieving ability have been: the separation of nitrogen and oxygen in air on the basis of difference of diffusion rates of these gases in small micropores; control of fragrance of winery products where only small molecules are removed.

Activated carbon fiber: Another solid is the synthetic fibers such as phenolic resin (Kynol R), polyacrylic resin (PAN), and viscose rayon. Most ACFs have fiber diameter of 7 to 15 pm, which is even smaller than powdered activated carbon.

GAS—Solid Adsorption Enthalpy (Calorimetric Methods)

The gas adsorption on a solid can be studied by various suitable apparatuses as regards the change in temperature. There are many calorimeters commercially available that can provide this information. The principal of these procedures is to measure the heats of gas adsorption on solids.

The experiment is carried out by measuring the change in temperature of the solid. Further, it is worth recalling that not only the magnitude of the heat evolved during adsorption but also its variation upon increasing coverage may reveal useful information concerning the type of adsorbate/surface sites bonding, and its evolution according to the surface heterogeneity.

As pointed out elsewhere (Chapter 3), the surface of a real solid material is in general characterized by a structural and/or a chemical heterogeneity of the sites, owing to the presence of either structural defects and/or (hetero)atoms in different oxidation states. Another kind of surface heterogeneity, originated by the presence of lateral interactions among adsorbed species, is the so-called induced heterogeneity.

The quantity the heat measured (calorimetric cells) represents the enthalpy change associated to the adsorption. This result applies to adsorption processes performed in a *gas-solid open* system through the admission of the adsorptive on the solid material kept isothermally within a heat-flow micro-calorimeter consisting of two cells in opposition. These micro-calorimeters are highly sensitive.

The thermodynamics of gas—solid adsorption has been investigated by using calorimeters (Adamson & Gast, 1997; Keller et al., 1992; Lowell et al., 2006; Bolis et al., 1990; Auroux, 2013).

MICROCALORIMETRY FOR GAS ADSORPTIONON SOLIDS: There are commercial apparatuses available for measuring gas adsorption on solids. The stepwise adsorption microcalorimetry technique is useful for providing quantitative data in surface chemistry studies. These data provide information about the nature of the adsorption process, physical or chemical. The solid surfaces have been extensively investigated by various spectroscopic techniques (infrared (IR) and Raman, UV-vis, NMR, XPS, EXAFS-XANES) (Ertl, 2003).

DIVERSE GAS ADSOPTON PRINCIPLES

The thermodynamics of gas adsorption on solids has been investigated by various methods such as calorimetry.

The heat (enthalpy) of gas adsorption on solids has been measured by using different types of calorimeters. The heat evolution process when a gas or a fluid interacts with the solid surface is related to the nature and energy of the adsorbed species/surface atoms interactions. The surface forces determine the thermodynamic process (exothermic or endothermic reaction). The enthalpy may be positive or negative. This provides a very useful information as regards the mechanism of the property. Further, the knowledge of the energetics of chemical and physical events responsible for the process as well as the assessment of the associated thermodynamic parameters contributes to a molecular understanding of the phenomena taking place at any kind of interfaces (Auroux, 2013; Chattoraj & Birdi, 1984; Bolis et al., 1998; Birdi, 2002).

In some cases, these studies have been correlated with other investigations such as spectroscopic or atomic microscopic studies.

The quantity enthalpy (q) measured is analyzed as follows. The entropy, ΔS_{ads}, and enthalpy, ΔH_{ads}, are estimated as follows.

$$\Delta S_{ads} = q + R - RT \ln(p^{1/2}) \qquad (B.1)$$

The expression on the right-hand side of the formula gives the enthalpic and the pressure contributions to the standard entropy of adsorption. The quantity enthalpy is expressed as

$$\Delta H_{ads} = q / T + R \qquad (B.2)$$

is obtained from the calorimetric data, whereas the free energy term

$$\Delta G_{ads} = -RT \ln (p_{gas}^{1/2}) \qquad (B.3)$$

is obtained from the adsorption isotherms. The quantity $p_{gas}^{1/2}$ is the equilibrium pressure when half of the surface is covered with gas.

The data of CO adsorption on Na—MFI and K—MFI were analyzed (at T = 673 K). The calorimetric heat of adsorption was ca. 35 and 28 kJ mol^{-1}

for Na—MFI and K—MFI, respectively. The half-coverage equilibrium pressure (obtained by the adsorption isotherms) were $p^{1/2}$ = 200 Torr for Na–MFI and 850 Torr for K–MFI. The magnitude of standard adsorption entropy was

$$\Delta S_a^\circ = -151 \text{ J mol}^{-1} \text{ K}^{-1} \text{ (for Na—MFI)}$$

and -140 J mol^{-1} K^{-1} (for K–MFI)

As an example, for the CO adsorption data, it was found that the decrease of entropy for CO adsorbed at a polar surface through electrostatic forces was slightly larger than for the a specific interaction of Ar atoms adsorbed at an apolar surface. This is as one would expect from physical interactions. The magnitude of entropy of adsorption, S_{ads}, of a gas will be expected to be related to the energy of adsorption. This was found from the data of adsorption of CO on MFI. The change (loss) of entropy for CO adsorbed on Na–MFI was found to be larger than on K–MFI. This was in accord with the higher energy of adsorption of CO on Na$^+$ than on K$^+$ sites.

There are also reported gas adsorption studies where more than one kind of adsorbed species may be present. This was found in the CO adsorption on TiO_2. It was found that in the TiO^2 (dehydrated) surface, CO adsorption showed two adspecies.

This was ascribed to the existence of two distinct IR bands located at v_{CO} = 2188 and 2206 cm^{-1} (Bolis et al., 1989). These analyses showed that two adspecies were formed on two different Lewis acidic sites made up of structurally different *cus* Ti^{4+} cations (species A and B). These A and B species showed the following spectroscopic and energetic properties:

Species A (v_{CO} = 2188 cm^{-1}) were formed at the 5th-coord Ti^{4+} cations typically exposed at the flat planes of anatase nanocrystals;

Species B (v_{CO} = 2206 cm^{-1}) were formed at the 4-coord Ti^{4+} cations, which are exposed at the steps, corners, and kinks of the flat planes (Rouqerol et al., 1998). It was found that species A's $v_{CO \text{ frequency}}$ moved from 2188 down to 2184 cm^{-1}.

The measurement of the heat of adsorption by a suitable calorimeter is the most useful method for evaluating and estimating the energy of gas adsorption (physical or chemical).

DIFFERENT CALORIMETERS USED FOR GAS ADSORPTION

Tian-Calvet heat-flow microcalorimeters are an example of high-sensitivity apparatuses that are suitably adapted to the study of gas—solid interactions when connected to sensitive volumetric systems. Volumetric-calorimetric data reported in the following were measured by means of either a C-80 or MS standard heat-flow microcalorimeter (both by Setaram, France).

In these studies, both the integral heats and adsorbed amounts were measured. Two identical calorimetric vessels, one containing the sample under investigation and the other (usually empty) serving as reference element, were connected in opposition.

In some studies, a stepwise procedure was used (Bolis et al., 1999). Small successive doses of the adsorptive were admitted and left in contact with the adsorbent until the thermal equilibrium was observed. After a definite amount of gas was introduced in the calorimeter, the amount of heat, ΔQ_{int} was measured, while the adsorbed amount of gas, Δn_{ads} was measured separately. Integral heats normalized to the adsorbed amounts are referred to as the integral molar heat of adsorption at the given equilibrium pressure p_{gas}:

$$\left(q_{mol}\right)p_{gas} = \left(Q_{int}/n_{ads}\right)p_{gas} \tag{B.4}$$

in kJ mol^{-1}. The quantity q_{mol} refers to an intrinsically average value and is related to different thermal contributions from the interactions between the gas molecules. In the case of non-interacting binding sites, the heat of adsorption is constant, regardless of the degree of surface coverage. Hence, the integral heats curve is a straight line through the origin, and the slope is equal to the differential heat of adsorption (q_{dif}). It is necessary to understand the gas adsorption process, as regards the dependence of degree of coverage and molecular interactions. This information is obtained from heats of adsorption data as a function of the degree of adsorption. The magnitude of the heat evolved during adsorption, which depends on the nature of the adsorbate/surface sites' bonding, varies upon increasing coverage as a consequence of the presence of either a heterogeneous distribution of surface sites or lateral interactions among adsorbed species.

For example: The adsorption of water on H–BEA and BEA-zeolites has been analyzed.

These studies were carried out as a function of water adsorbed amounts or water equilibrium pressure.

From these studies it was concluded that it will be reasonable to expect that at the $Si(OH)^+Al^-$ sites water (H_2O) molecules are adsorbed (with strong hydrogen bonds).

The heats of adsorption started from a quite high zero-coverage value ($h_{ad} \approx 160$ kJ mol^{-1}), which is compatible with a chemisorption process, either the protonation of H_2O at the (Brønsted) acidic site or the strong oxygen-down coordination at the (Lewis) acidic sites. For each H_2O molecule adsorbed per Al atom, on average, the heat values were found to be in the range of $160 < h_{diff} < 80$ kJ mol^{-1}, whereas for the second to fourth H_2O adsorbed molecules in the $80 < h_{diff} < 60$ kJ mol^{-1} range.

In the all-silica BEA specimen, the zero-coverage heats of adsorption were much lower than for H—BEA ($h_{ad} \approx 70$ vs.160 kJ mol^{-1}, respectively).

These data thus suggested that the all-silica zeolite behaves as a non-hydrophobic surface.

Furthermore, these data showed that for adsorption leading to the same residual pressure in different runs and/or on different samples, the values of q_{diff} were almost the same. A constant value for the differential heat was obtained: $h_{dif} \approx 35$ kJ mol^{-1} for Na—MFI and ≈ 28 kJ mol^{-1} for K—MFI. The linear fit of the integral heat curves seemed the most realistic, in spite of the fact that in both cases at very low and at high coverage the middle points of the experimental histogram deviated from the constant value. In fact, the low-coverage heterogeneity was due to the presence of a few defective centers (1–2% of the total active sites) interacting with CO more strongly than the alkaline metal cations.

For example: The enthalpy of adsorption of ammonia (NH_3) have been reported in literature (Bolis et al., 1999). The system that was studied was:

NH_3 adsorption on H—MFI and all-silica MFI zeolites. The data for enthalpy of the reversible adsorption of NH_3 on different MFI—Silicalite (Sil–A, Sil–B, Sil–C) and on a perfect (i.e., defect-free) MFI—Silicalite (Sil–D) have been studied. The values of heats of adsorption started from a quite high zero-coverage value ($h_{ad} \approx 160$ kJ mol^{-1}), which is compatible with a chemisorption process, either the protonation of H_2O at the Brønsted acidic site or the strong oxygen-down coordination at the Lewis acidic sites.

For one H_2O molecule adsorbed per Al atom, on average, the heat values were found to be in the range of $160 < h_{dif} < 80$ kJ mol^{-1} range, whereas the second to fourth H_2O adsorbed molecules were found in the $80 < h_{dif} < 60$ kJ mol^{-1} range.

It has been explained that the volumetric technique is more accurate at low pressure because almost all of the metered dose is adsorbed. The gravimetric technique has the disadvantage at low pressure in that the amount adsorbed is the difference of two nearly equal numbers. At high pressure, the volumetric technique gives the amount adsorbed as the sum of a large number of doses with an associated cumulative error. The gravimetric technique is more accurate at high pressure because the measured amount adsorbed is referenced to the weight of the adsorbent in a vacuum.

Volumetric method. The volumetric technique introduces a known mass of adsorbent into a sample cell of calibrated volume.

METHODS OF GAS ADSORPTION ON SOLIDS

In the literature, one finds various methods that have been used to study the process of gas adsorption on solid surfaces. This process has been studied for almost a century. Further, it is also one of the most developing processes. This arises from the fact that gas adsorption plays a very important role in many important industries (such as gas purification and separation technology, catalysis, flue gas treatment, water purification, pollution control, gas sensor technology).

The different methods used depend on the system under investigation. There are also many commercially available instruments that are designed for any specific process under investigation.

Gravimetric method. In the gravimetric method the amount of substance adsorbed on another phase is measured. A mass of solid adsorbent is loaded into a container attached to a microbalance. Following desorption of the solid using high temperature and vacuum, the system is brought to a specified temperature and gas is admitted to the sample cell. After adsorption is complete, at equilibrium, the temperature (T) and pressure (P) are measured and the adsorption is determined from the weight of the solid + adsorbed gas. The weight of gas adsorbed is equal to the weight of the container with the solid minus its degassed tare weight under vacuum.

Volumetric method: In this procedure the change in volume of the system is measured. The change in volume corresponds to the gas adsorbed (Adamson & Gast, 1997; Keller et al., 1992; Birdi, 2002).

Porous solids: In the case of porous solids, the gas adsorption needs a different approach.

The pore volume of the solid (v_p) has been estimated from the amount of an inert gas (helium) adsorption in the pores at ambient temperature (Keller et al., 1992).

As explained above, every solid exhibits a specific surface property: area/gram.

This is the quantity that is essential in the analyses of all gas adsorption data on solids. A general procedure used is to determine the adsorption of an inert gas, such as helium, on the solid (Myers, 1989; Adamson & Gast, 1997).

GAS SENSOR SOLID MATERIALS

The gas adsorption on a solid gives rise to a change in the physical properties of the solid. The latter observation thus leads to the possibility of using the change in solid characteristic as the gas sensor (Korotcenkov, 2013). Some measuring techniques show that these changes in the solid properties are related to the gas adsorption. Hence, this observation can be used as a sensor for the gas.

It is found that it is not easy to characterize an ideal sensing material. One finds a whole range of gas sensors in industrial applications. This shows the important application of the gas—solid adsorption process. For instance, one finds sensors that can detect a variety of gases: CO_2, CO, H_2, NO_2, NH_3, Cl_2 (Korotcenkov, 2013).

This observation shows that gas adsorption on solid creates a new solid surface with different physical properties.

CO_2 CAPTURE FROM AIR

There are literature investigations that analyze the technology that could capture CO_2 from air (400 ppm). It is thus useful because it can be applied at all suitable sites. Of course, the main challenge for an efficient process is that the concentration of CO_2 is relatively low (Dubey et al., 2002).

CRYOGENIC DISTILLATION OF CO_2

In any CCS process, the aim is to produce almost pure CO_2. Mainly, this procedure leads to either usage of CO_2 in industry or storage in geological reservoirs.

There exists another method by which CO_2 can be extracted from flue gas. This is based on cryogenic distillation technology.

This process consists of a gas separated from flue gas by using distillation at very low temperature and high pressure, which is similar to other conventional distillation processes except it is used to separate components of gaseous mixture (owing to their different boiling points) instead of liquid. For CO_2 separation, flue gas-containing CO_2 is cooled to desublimate temperature (−100 to −135°C), and then solidified CO_2 is separated from other tight gases and compressed to a high pressure of 100–200 atmospheric pressure. The amount of CO_2 recovered can reach 90–95% of the flue gas. Since the distillation is conducted at extremely low temperature and high pressure, it is an energy intensive process estimated to require 600–660 kWh per ton of CO_2 recovered in liquid form.

CO_2 HYDRATE FORMATION AND CCS

This procedure is based on a specific property of ice and hydrate formation.

All substances when in solid state have higher density than in the liquid state. Ice floats on water, which is an exception.

It is known that certain gas molecules form hydrates with ice (CH_4, CO_2, Cl_2) (Tanford, 1980; Birdi, 2016). There is another new procedure reported for CCS. This is based on the formation of a complex (hydrate) between water molecules in ice and a gas molecule.

Physical Properties of Carbon Dioxide (CO_2) Properties

A short description of some important chemical properties is useful at this stage. The different chemical characteristics are delineated under a separate section. This is useful in understanding the different processes mentioned in the text.

CO_2 is a gas at standard temperature and pressure (1 atm). CO_2 is only found in the Earth's atmosphere (currently ca. 410 ppm). There has been observed an increase in CO_2 concentration in the atmosphere over the last decades. In the molecule of CO_2, the distance between C and O is 116.3 pm (which corresponds with double bond). Hence, the CO_2 molecule is planar (O=C=O) and linear in space. The molecule has no electrical dipole. Some physical properties of CO_2 are given in the following.

Physical properties of CO_2

Molecular weight	44.03 g/mole
Color	Colorless and odorless
Density/pressure/temperature	1.562 gm/ml/Solid/1 atm/−78.5°C
	0.77 gm/ml/Liquid/56 atm/20°C
	1.977 gm/liter/Gas/1 atm/0°C

Melting point	−78 °C (194.7 K) (sublimation)
Boiling point	−57 °C (216.6 KJ/t.185 bar)
Solubility in water	1.45 gm/liter at 25 °C, 100 kPa (1 atm)
Refractive index (nD)	1.1120
Viscosity	0.07 cP at −78 °C
Molecular shape (O = C = O)	Linear
Dipole moment	zero

Appendix C

The purpose of this Appendix is to add some more basic information on the CCS technology described in Chapter 4. CCS technology is expanding rapidly, and there are many different aspects that need to be addressed. Some additional data are mentioned here for those readers who may be interested in pursuing further studies in CCS technology.

GLOBAL AVERAGE TEMPERATURE DATA

The average temperature of the Earth is not easy to define. The general procedure is that some chosen sites around the world are used for monitoring temperature.

In the following text, some typical temperature data are given (Mathews, 2009), although there are very few and they only point to the complexity of these data. The data include land and sea data during the year 2014 (Berkeley Earth, 2014). Some important features may be mentioned.

Ranking order is thus (year/degree):

1. 2014, 0.6;
2. 2010, 0.6;
3. 2005, 0.5;
4. 2007, 0.5;
5. 2006, 0.5;
6. 2013, 0.5;
7. 2009, 0.5;
8. 2002, 0.5;
9. 1998, 0.5;
10. 2003, 0.5.

SUN AND THE TEMPERATURE ON EARTH

The Sun is 100 million miles away from the Earth. It is very large as compared to Earth (radii of Sun and Earth, respectively: 695,700 km and 6370 km). The solar energy reaching the Earth varies with time. It is found that there exists an 11-year solar activity cycle.

The Sun provides the primary source of energy (heat) driving Earth's climate system, but its variations have played a minor role in the climate changes observed in recent decades. Direct satellite measurements since the late 1970s show no net increase in the Sun's output, while at the same time global surface temperatures have increased. For earlier periods, solar changes are less certain because they are inferred from indirect sources—including the number of sunspots and the abundance of certain forms (isotopes) of carbon or beryllium atoms, whose production rates in Earth's atmosphere are influenced by variations in the Sun spots.

Solar activity cycle:

Data shows that the 11-year solar cycle, during which the Sun's energy output varies by roughly 0.1%, can influence ozone concentrations, temperatures, and winds in the stratosphere (the layer of the atmosphere above the troposphere, typically from 12 to 50 km, depending on latitude and season).

These stratospheric changes may have a small effect on surface climate over the 11-year cycle. However, the available evidence does not indicate pronounced long-term changes in the Sun's output over the past century, during which time human-induced increases in CO_2 concentrations have been the dominant influence on the long-term global surface temperature increase. Further evidence that current warming is not a result of solar changes can be found in the temperature trends at different altitudes in the atmosphere.

The measurements of the Sun's energy incident on Earth do not show any increase in solar flares during the past 30 years. The data show only small periodic amplitude variations associated with the Sun's 11-year cycle. Some results from mathematical/physical models of the climate system showed that human-induced increases in CO_2 would be expected to lead to gradual warming of the lower atmosphere (the troposphere) and cooling of higher levels of the atmosphere (the stratosphere). In contrast, increases in the Sun's output would warm both the troposphere and the full vertical extent of the stratosphere. At that time, there was insufficient analytical data to test this prediction, but temperature measurements from weather balloons and satellites have since concerned these early forecasts. It is now known that the observed pattern of tropospheric warming and stratospheric cooling over the past 30 to 40 years is in general consistent with computer model simulations that include increases in CO_2 and decreases in stratospheric ozone, each caused by human activities. These data are not consistent with purely natural changes in the Sun's energy output, volcanic activity, or natural climate variations such as El Niño and La Niña.

Despite this agreement between the global-scale patterns of modeled and observed atmospheric temperature change, there are still some differences. The most noticeable differences are in the tropical troposphere, where models currently show more warming than has been observed, and in the Arctic, where the observed warming of the troposphere is greater than in most models. The observed data on the warming in the lower atmosphere and cooling in the upper atmosphere provides some useful information on the reasons of climate change.

CLIMATE CHANGE (GENERAL REMARKS)

In this Appendix a short description is given to describe some studies as related to the climate (as regards temperatures) of the Earth. The past observations indicate that the climate on Earth is variable over short and long timescales. The most obvious is the ice age, which led to the melting of ice around Northern Europe (some 15,000 years ago). The largest global-scale climate variations in Earth's recent geological past are the ice age cycles), which are cold glacial periods followed by shorter warm periods. The last few of these natural cycles have recurred roughly every 100,000 years. They are mainly paced by slow changes in Earth's orbit, which alter the way the Sun's energy is distributed with latitude and by season on Earth. These changes alone are not sufficient to cause the observed magnitude of change in the temperature, nor to act on the whole Earth. Instead they lead to changes in the extent of ice sheets and in the abundance of CO_2 and other greenhouse gases, which amplify the initial temperature change and complete the global transition from warm to cold or vice versa.

Recent estimates of the increase in global average temperature since the end of the last ice age are 4 to 5°C (7 to 9°F). That change occurred over a period of about 7000 years, starting 18,000 years ago. CO_2 has risen by 40% in only the past 200 years, contributing to human alteration of the planet's energy budget that has so far warmed the Earth by about 0.8°C (1.4°F).

Measurements of composition of air in ice cores show that for the past 800,000 years, up until the 20th century, the atmospheric CO_2 concentration stayed within the range 170 to 300 parts per million (ppm), making the recent rapid rise to nearly 400 ppm over 200 years particularly noticeable. During the glacial cycles of the past 800,000 years both CO_2 and methane have acted as important amplifiers of the climate changes triggered by variations in Earth's the Sun.

Furthermore, since the very warm year 1998, with the strong 1997–98 El Niño, the increase in average surface temperature has slowed down. However, despite the slower rate of warming, the 2000s were warmer than the 1990s. Decades of slow warming as well as decades of accelerated warming have occurred naturally in the climate system. Also, decades that are cold or warm compared to the long-term trend are seen in the observations of the past 150 years and also captured by climate models.

Most significantly, more than 90% of the heat added to the Earth is absorbed by the oceans and penetrates slowly into deep water. Owing to the lack of mixing, the system is thus at a pseudo equilibrium. A faster rate of heat penetration into the deeper ocean will slow the warming seen at the surface and in the atmosphere but by itself will not change the long-term warming that will occur from a given amount of CO_2. For example, recent studies show that some heat comes out of the ocean into the atmosphere during warm El Niño events, and more heat penetrates to ocean depths in cold La Niñas. Such changes occur repeatedly over timescales of decades and longer. An example is the major El Niño event in 1997–98 when the globally averaged air temperature soared to the highest level in the 20th century as the oceans lost heat to the atmosphere, mainly by evaporation. Recent studies have also pointed to a number of other small cooling sequences over the past decade or so. These include a relatively quiet period of solar activity and a measured increase in the amount of aerosols (reactive particles) in the atmosphere owing to the cumulative effects of a succession of small volcanic eruptions.

Studies showed that while Arctic sea ice is decreasing, Antarctic sea ice is not.

Sea ice in the Arctic has decreased dramatically since the late 1970s, particularly in summer and autumn. Since the satellite record began in

1978 (providing for the first time a complete and continuous aerial coverage of the Arctic), the yearly minimum Arctic sea ice extent (which occurs in early to mid-September) has decreased by more than 40%.

The effect of winds on sea ice is also a large factor. The changes in wind direction and in the ocean seem to be dominating the patterns of climate around Antarctica.

DIVERSE CARBON DIOXIDE (CO_2) SOURCES AND SINKS

Photosynthesis, Respiration, and CO_2 on Earth

Life on Earth is mainly dependent on the interaction between the Sun, air, and water, besides other phenomena. In particular, sunlight is responsible for many life-dependent phenomena. The main effects of sunlight are heat and photosynthesis. One of the most important natural phenomena on Earth is photosynthesis (Rabinowitch & Govindjee, 1969). This subject is out of scope of the present context, but only a short description is given here. It is obvious that one cannot estimate accurately the total yield of photosynthesis on the Earth's surface. The yield in ocean plants is very large. Some data of photosynthesis yield for different vegetation types are given in Table C.1 (Rabinowitch & Govindjee, 1969).

It is suggested that the Sun and Earth were born simultaneously (Manuel, 2009). The Sun is known to produce heat from hydrogen fusion (fusion of hydrogen (75%) to helium (24%)).

The Sun consists mainly of hydrogen (H_2; 90%). The variations in solar activity, for example, cycles of solar eruptions, cosmic rays, sunspots, and variations in magnetic parity (Manuel, 2009; IPCC, 2007; McCormmach, 1970; Akasofu, 2007; Jose, 1965).

TABLE C.1 Typical Photosynthesis Yields of Carbon Dioxide Captured (Tons Carbon/ Year) Into Organic Matter

Vegetation	Area (10^6 km²)	Tons of C/km²	Total Yield (10^9 tons C/year)
On land			
Forests	44	250	11
Grassland	31	35	1.1
Farmland	27	150	4
Desert	47	5	0.22
Total on land	149	...	16

The process of photosynthesis converts CO_2 (in conjunction with H_2O) in air to all kinds of plants and food for mankind. This complicated process produces the most essential building block, glucose ($C_6H_{12}O_6$) for the formation of cellulose and carbohydrates. The main photosynthesis reaction is

$$6\ CO_2 + 6H_2O + \text{sunlight} ========== C_6H_{12}O_6 + 6\ O_2$$

Catalyst

In this very important life-sustaining reaction, the various chemicals are acquired from different sources.

CO_2 is provided by air (with a current concentration ca. 400 ppm);

Water is present as moisture;

Or is obtained from soil.

The green color in plants is the chlorophyll molecule, which is the catalyst of the reaction. The green color in plants is related to chlorophyll molecule.

LUNG FUNCTION AND HUMAN METABOLISM (CO_2 CYCLE)

Life on Earth, especially mankind, is dependent on some very essential needs; one of these is food. Food (such as corn, rice, fruits, etc.) is made by photosynthesis of CO_2. Food is used for growth through metabolism.

The metabolism of mankind and many other living species is simply

FOOD INTAKE—METABOLISM—PRODUCTION OF CARBON DIOXIDE AND OTHER WASTE PRODUCTS

However, food is produced through photosynthesis (CO_2 from air). It is thus seen that the metabolism is CO_2 neutral (almost).

The function of the respiratory system of humans (and other living species) shows that in the lungs different gases (from air) are exchanged (i.e., from blood to air in the inhaled lungs) as a result of metabolic processes. Breathing brings the oxygen (O_2) in the air (28%) into the lungs and into close contact with the blood, which absorbs it and carries it to all parts of the body. At the same time, the blood delivers CO_2, which is carried out of the lungs when air is breathed out (Nienstedt et al., 1992).

Human metabolism:

Blood flow into lungs = Hemoglobin-CO_2

(Exchange of CO_2 with O_2 in lungs)

Blood flow out from lungs = hemoglobin-O_2

The composition of the inspired and expired air is found to be of following composition:

	Inspired Air	Expired Air
Oxygen	21%	16%
CO_2	0.04%	4%
Nitrogen	78.0%	78%

The principal function of the lung is to exchange oxygen (absorb from air) and CO_2 (desorb to air) between blood and inspired air. In the lungs, hemoglobin carries CO_2 to the lungs and releases it. Simultaneously, hemoglobin absorbs oxygen from air and carries it inside the body to be used. Blood vessels in the lungs have a structure similar to the bronchial tree. The pulmonary artery carries de-oxygenated blood from the right ventricle to the lungs. The pulmonary artery branches first to the left and right lung and branches further down to the capillary level. The pulmonary veins carry blood from the lungs to the left atrium of the heart. They have inverse structure compared to the pulmonary arteries, starting on the capillary level and reaching the main pulmonary vein, which leads to the heart.

The hemoglobin molecule carries CO_2 to the lungs and exchanges it with oxygen (from inhaled air). However, if CO_2 concentration in the air is over 3%, then breathing becomes difficult (as may happen in closed environments, such as mines, etc.). In fact, above 14%, CO_2 is fatal to human life. This shows that the food—metabolism cycle is almost neutral, excepting energy used for food production and transport.

COST OF CO_2 CAPTURE

Cost of CO_2 Capture and Storage:

Although the economical aspects of CCS technology are out of the scope of this book, a very short mention is provided. The different technologies

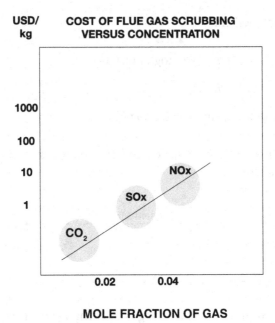

FIGURE C.1 A plot of various gas-scrubbing processes in which cost increases with decreasing concentration. Concentration is approximately 12 mol% CO_2 (Lightfoot & Cockrem, 1987).

that might be suitable for CCS will of course be dependent on the individual physical structures and background of the power plant (Bhown & Freeman, 2011; Matthews & Caldeira, 2008). Additionally, the geographic situation will also be different.

The economy will be expected to be dependent on different factors. The costs of separation may be divided into technical versus non-technical costs.

In the case of non-technical costs, one may include account depreciation and return on investment, interest rate, labor, etc. Costs associated with the technology may include the expenses related to the equipment, chemicals used, power consumption, power cost, etc. For instance, if sustainable power sources are used (such as solar energy, wind energy), then the costs will be much different.

Additional factors that may affect the cost of CO_2 capture may also include the type of power plant and capture technology. For instance, will one use an existing power plant or will the CCS technology be applied to a new plant? These process design and variables are much too complex and need to be investigated. Other factors that will also need to be considered are capture capacity, capture rate, and CO_2 concentration (flue gas: in and out).

It is obvious that in general, the carbon capture cost will be related to the concentration of CO_2 in the flue gas (Moniz & Tinker, 2010). The higher the starting concentration of CO_2, the lower the average cost of capture. A plot of (Figure C.1) cost increase with increasingly dilute CO_2 flue gas has been reported. This kind of plot is called the Sherwood correlation.

Bibliography

Abbas, S., and Nordholm, S., *J. Colloid Interface Sci.*, 166, 481, 1994.

Aboudheir, A., Tontiwachwuthikul, P., Chakma, A., and Idem, R., *Chemical Engineering Science*, 5195, 58, 2003.

Abunowara, M., Bustam, M.A., Sufian, S., Eldemerdash, U., *Procedia Engineering*, 148, 600, 2016.

Adam, N.K., *The Physics and Chemistry of Surfaces*, Clarendon Press, Oxford, 1930.

Adamson, A.W., and Gast, A.P., *Physical Chemistry of Surfaces*, 6th Edition, Wiley Interscience, New York, 1997.

Akasofu, S.I., *Exploring the Secrets of the Aurora*, Springer Publ. Co., New York, 2007.

Albo, A., Luis, P., and Irabin, A., *Ind. Eng. Chem. Res.*, 11045, 49, 2010.

Alie, C., Backham, L., Croiset, E., and Douglas, P.L., *Energy Conversion and Management*, 475, 46, 2005.

Amidon, G.L., Yalkowsky, H., and Leung, S.J., *J. Pharmaceutical Sci.*, 3225, 63, 1974.

Aronu, U.E., Svendsen, H.F., and Hoff, K.A., *International Journal of Greenhouse Control*, 771, 4, 2010.

Astarita, G., *Chemical Engineering Science*, 202, 16, 1961.

Attard, G., and Barnes, C., *Surfaces*, Oxford Chemistry Primes, No. 59, Oxford Science Publications, Oxford, 1998.

Auroux, A. (ed.), *Calorimetry and Thermal Methods in Catalysis, 3 Springer Series in Materials Science 154*, Springer-Verlag, Berlin, Heidelberg, 2013.

Aveyard, R., *J. Colloid Interface Sci.*, 52, 621, 1975.

Aveyard, R., and Briscoe, B.J., *J. Chem. Soc. Faraday Trans.*, 68, 478, 1972.

Aveyard, R., and Hayden, D.A., *An Introduction to Principles of Surface Chemistry*, Cambridge University Press, Cambridge, UK, 1973.

Avnir, D. (ed.), *The Fractal Approach to Heterogenous Chemistry*, Wiley, New York, 1989.

Back, D.D., *Journal of Chemical & Engineering Data*, 41, 446, 1996.

Bakker, Z., *Physik. Chem.*, 89, 1, 1815.

Barrer, V.R.M., *Zeolites and Clay minerals as Sorbents and sieves*, Academic Press, New York, 1978.

Bartos, B., Freund, H.J., Kuhlenbeck, H., Neumann, M., Lindner, H., and Mfiller, K., *Surface Science*, 59, 179, 1987.

Beising, R., *Climate Change and the Power Industry*, VGB Power Technology, March 2007.

Benadda, B., Prost, M., Ismaily, S., Bressat, R., and Otterbein, M., *Chemical Engineering and Processing*, 55, 33, 1994.

Berkeley Earth, Berkeley, www.BerkeleyEarth.org, 2014.

Bhatia, A.B., March, N.H., and Sutton, J., *J. Chem. Phys.*, 2258, 69, 1978.

Bhown, A.B., and Freeman, B.C., *Environ. Sci. Technol.*, 8624, 45, 2011.

Birdi, K.S., The Hydrophobic Effect, *Trans. Faraday Soc.*, 17, 194, 1982.

Birdi, K.S., *Lipid and Biopolymer Monolayers at Liquid Interfaces*, Plenum Press, New York, 1989.

Birdi, K.S., *Fractals (In Chemistry, Geochemistry and Biophysics*, Plenum Press, New York, 1993.

Birdi, K.S. (ed.), *Handbook of Surface & Colloid Chemistry*, CRC Press, Boca Raton, 1997.

Birdi, K.S., *Self-Assembly Monolayer (SAM) Structures*, Plenum Press, New York, 1999.

Birdi, K.S. (ed.), *Handbook of Surface & Colloid Chemistry*, 2nd Edition, CRC Press, Boca Raton, 2002.

Birdi, K.S., *Scanning Probe Microscopes (SPM)*, CRC Press, Boca Raton, 2003.

Birdi, K.S. (ed.), *Handbook of Surface & Colloid Chemistry*, 3rd Edition, CRC Press, Boca Raton, 2009.

Birdi, K.S., (ed.), *Introduction to Electrical Interfacial Phenomena*, CRC Press, Boca Raton, 2010.

Birdi, K.S., *Surface Chemistry Essentials*, CRC Press, Boca Raton, 2014.

Birdi, K.S. (ed.), *Handbook of Surface & Colloid Chemistry*, 2nd Edition, CRC Press, Boca Raton, 2003.

Birdi, K.S., *Surface Chemistry & Geochemistry of Hydraulic Fracturing*, CRC Press, Boca Raton, 2017.

Birdi, K.S., and Vu, D.T., *Adhes. Sci. Technol.*, 7, 485, 1993.

Birdi, K.S, Vu, D.T., and Winter, A., *Proc. 4th Euro. Symp. Enhanced Oil Recovery*, Hamburg, Germany, October 1987.

Birdi, K.S., Vu, D.T., and Winter, A., *J. Phys. Chem.*, 93, 3702, 1989.

Bishnoi, S., and Rochelle, G.T., *Chem. Eng. Sci.*, 55, 5531, 2000.

Bliem, I.L., Kosak, R., Erneczky, L., Novotny, Z., Gamba, O., Fobes, D., Mao, Z., Schmid, M., Blaha, P., Diebold, U., and Parkinson, G.S., *ACS Nano*, 7531, 8, 2014.

BNEF (Bloomberg New Energy Finance), *Clean Energy Investment Trends*, BNEF, London, 2016.

Bolis, V., in *Calorimetry & Thermal Methods in Catalysis*, ed. by A. Auroux, Springer, New York, 2013.

Bolis, V., Barbalia, A., Bordiga, S., Lamberti, C., and Zecchina, *J. Phys. Chem. B*, 9970, 108, 2004.

Bolis, V., Bordiga, S., Lamberyi, C., Zecchina, A., Carati, A, Rivetti, F., Spano, G., and Petrini, G., *Langmuir*, 5753, 15, 1999.

Bolis, V., Busco, C., Aina, V., Morterra, C., and Ugliengo, P., *J. Phys. Chem. C*, 16879, 112, 2008.

Bolis, V., Cerrato, G., Magnacca, G., and Morterra, C., *Biochem. Biochim. Acta*, 63, 31, 1998.

Bolis, V., Fubini, B., Garrone, E., Giamello, C., Morterra, in: *Structure and Reactivity of Surfaces*, Vol. 48, ed. by C. Morterra, A. Zecchina, G. Costa, Elsevier Sci. Publ. B.V., Amsterdam, 159, 1989.

Bolis, V., Fubini, B., Marchese, L., and Martra, G., *J. Chem. Soc., Faraday Trans.*, 497, 87, 1991.

Bolis, V., Morterra, C., Volante, M., Orio, L., and Fubini, B., *Langmuir*, 695, 6, 1990.

Bonenfant, D., Kharoune, M., Niquette, P., Mimeault, M., and Hausler, R., *Sci. Technol. Adv. Mater.*, 13007, 9, 2008.

Breck, D.W., *Zeolits Molecular Sieves*, vol. 4, John Wiley, New York, 1974.

Brilman, D.W.F., Goldschmidt, M.J.V., Versteeg, G.F., and van Swaaij, W.P.M., *Chemical Engineering Science*, 2793, 55, 2000.

Britt, D., Furukawa, H., Wang, B., Glover, T.G., and Yaghi, O.M., *PNAS*, 20637, 106, December 8, 2009.

Bruining, W.J., Joosten, G.E.H., Beenackers, A.A.C.M., and Hofman, H., *Chemical Engineering Science*, 1873, 41,1986.

Brunauer, S., Emmett, P.H., and Teller, E., *J. Am. Chem. Soc.*, 309, 60, 1938.

Brunet, Schaub, R., Fedrigo, S., Monet, R., BuRet, J., and Harbich, W., *Surface Science*, 201, 512, 2002.

Bryngelsson, M., and Westermark, M., *Proceedings of the 18th International Conference on Efficiency, Cost, Optimization, Simulation and Environmental Impact of Energy Systems*, 703, 2005.

Burg, T.P., Godin, M., Knudsen, S.M., Shen, W., and Manalis, S.R., *Nature*, 446, 1066, 2007.

Burshtein, A.A., *Adv. Colloid Interface Sci.*, 11, 315, 1979.

Butt, H.J., Graf, K., and Kappl, M., *Physics and Chemistry of Interfaces*, J.W., Wiley, Indianapolis, 2006.

Caldecott, B., Lomax, G., and Workman, M., *Stranded Carbon Assets and Negative Emissions Technologies Working Paper*, Stranded Assets Programme, University of Oxford, 2015.

Callen, H.B., *Thermodynamics and an Introduction to Thermostatistics*, 2nd Edition, John Wiley & Sons, Inc., New York, NY, 1985.

Cambell, T.M., *Catalysis, at Surfaces*, Chapman & Hall, London, 1988.

Carbon Brief, *"Analysis: Only Five Years Left Before 1.5°C Budget Is Blown"*,

Carbon Capture and Storage at Imperial College London, Imperial College London, London, 2016.

Casket, S.R., Wong-Foy, A.G., Antek., A.J., *J. Am. Chem. Soc.*, 1087, 130, 2008.

CCS Programmes, Carbon Capture and Storage Association, London, 2008–2015.

CCS20 Years of Carbon Capture and Storage, OECD/IEA Accelerating Future Deployment, 2016.

CCSA (Carbon Capture and Storage Association), Lessons and Evidence Derived from the UK, 2016.

Cebrucean, D., *CO_2 Capture and Removal Technologies, Proceedings of the 2-nd International Conference on Thermal Engines*, Galati, Italy, 95, 2, 2007.

Cents, A.H.G., Brilman, D.W.F., and Versteeg, G.F., *Chemical Engineering Science*, 1075, 56, 2001.

Chattoraj, D.K., and Birdi, K.S., *Adsorption and the Gibbs Surface Excess*, Plenum Press, New York, 1984.

Chen, J., *Physics of Solar Energy*, John Wiley & Sons, New York, 2011.

Choudhary, V.R, and Mayadevi, S., *Zeolites*, 17, 501, 1996.

Cini, R., Loglio, G., and Ficalbi, A., *J. Colloid Interface Sci.*, 41, 287, 1972.

Cioci, F., *J. of Physical Chemistry*, 100, 17400, 1996.

Cornils, B., *Journal of Molecular Catalysis A: Chemical*, 1, 143, 1999.

Cosgrove, T., *Colloid Science*, Blackwell Publ., Oxford, 2005.

Cosgrove, T., Phipps, J., and Richardson, R.M., *Colloids Surf.*, 62, 199, 1992.

Cramer, C.J., and Truhlar, D.G., *Science*, 256, 213, 1992.

Cullinane, J.T., and Rochelle, G.T., *Ind. Eng. Chem. Res.*, 2531, 45, 2006.

Danckwerts, P.V., *Gas-liquid Reactions*, McGraw-Hill, New York, 1970.

Danckwerts, P.V., and Sharma, M.M. *Journal of Chem. Engineering Reviews Series*, No. 2, 244, 1966.

Dartois, E., *Astronomy & Astrophysics*, 869, 504, 2009.

Das, T.R., Bandopadhyay, A., Parthasarathy, R., and Kumar, R. *Chemical Engineering Science*, 209, 40, 1985.

Daschbach, J.L, Peden, B.M., Smith, R.S., and Kay, B.D., *The Journal of Chemical Physics*, 516, 120, 2004.

Davies, J.T., and Rideal, E.K., *Interfacial Phenomena*, Academic Press, New York, 1963.

Defay, R., Prigogine, I., Bellemans, A., and Everett, D.H., *Surface Tension and Adsorption*, Longmans, Green, London, 1966.

Dennis, J.D., *Investigation of Condensed and Early Stage Gas Phase Hypergolic Reactions (PhD thesis)*, Purdue University, USA, 2014.

Des Marais, D.J., Joseph, A., Nuth, J. A., Allamandola, L.J., Boss, A.P., Farmer, J.D., Hoehler, T.M., Jakosky, B.M., Meadows, V.S., Pohorille, A., Runnegar, B., and Spormann, A.M., *Astrobiology*, 715, 8, 2008.

Didas, S.A., Kulkarni, A.R., Sholl, D.S., and Jones, C.W., *Chem ASUS. Chem.*, 2058, 5, 2012.

Ding, X., De Rogatis, L., Vesselli, E., Baraldi, A., Comelli, G., Rosei, R., Savio, L., Vattuone, L., Rocca, M., Fomasiero, P., Ancilotto, F., Baldereschi, A., and Peressi, M., *Physical Review B*, 19, 76, 2007.

Dixon, T., McCoy, S.T., and Havercroft, I. *International Journal of Greenhouse Gas Control*, Elsevier, Amsterdam, 431, 40, 2015.

Dooley, J.J., Dahowski, R.T., and Davidson, C.L., *CO2-driven Enhanced Oil Recovery as a Stepping Stone to What?* Pacific Northwest National Laboratory, United States Department of Energy, 2010.

Douglas, M., *Ruthven, Principles of Adsorption and Adsorption Processes*, John Wiley & Sons, Inc., New York, NY, 1984.

Drelich, J., Lakskowski, J.S., and Mittal, K.L., *Apparent and Microscopic Contact Angles*, Intern. Science Publ., VSP, Leiden, 2002.

Dubey, M.E., Ziock, H., Rueff, G., Elliott, S., and Smith, W.S., *FUEL Chem.*, 81, 47, 2002.

Dubinin, M.M., *Chem Rev*, 235, 60, 1960.

Dunne, J.A., Myers, A.L., and Kofke, D.A.K, *Adsorption*, 41, 2, 1996.

Eckenfelder, W.W., and Barnhart, E.L. *AIChE. Journal*, 7, 631 (7), 1961.

EPA Handbook, *Exposure Factors Handbook*, U.S. Environ. Protection Agency, Washington, DC, 2011.

Ertl, G., in *Encyclopedia of Catalysis*, Vol. 8, 2nd Edition, ed. by J.T. Horvath, John Wiley & Sons, Hoboken, NJ, 2003.

Extrand, C.W., and Kumagai, Y., *J. Colloid Interface Sci.*, 191, 318, 1997.

Eyring, H., and Jhon, M.S., *Significant Liquid Structures*, John Wiley & Sons, London, 1969.

Falbe, J., *New Synthesis with Carbon Monoxide: Chapter V: Koch Reactions*, H. Bahrmann, Ed., Springer, Berlin, 1980.

Fanchi, J.R., and Fanchi, C.J., *Energy in the 21st Century*, World Scientific Publishing Co Inc., 2016.

Favre, E., *J. Membrane Sci.*, 294, 50, 2007.

Feder, J., *Fractals: Physics of Solids and Liquids*, Plenum Press, New York, 1988.

Feenstra, C.F.J., Mikunda, T., and Brunsting, S., *What Happened in Barendrecht? Case Study on the Planned Onshore Carbon Dioxide Storage in Barendrecht*, Energy Research Centre of the Netherlands, the Netherlands, 2010.

Feenstra, J.F., Burton, I., Smith, J.B., and Tol, R.S.J., *handbook on methods for climate change impact assessment and adaptation strategies*, UN Env. Program., Inst fdaa Envm. Studies, Vrije Univ., Amsterdam, Holland, Second Version, October 1908.

Felix, A., and Carroll, F.A., *Langmuir*, 16, 6689, 2000.

Fendler, J.H., and Fendler, E.J., *Catalysis in Micellar and Macromolecular Systems*, Academic Press, New York, 1975.

Ferst, P., Mehl, S., Arman, M.A., Schnler, M., Toghan, A., Laszlo, B., Lykhach, Y., Brummel, O., Lundgren, E., Knudsen, J., Hammer, L., Schneider, M.A., and Libuda, J., *The Journal of Physical Chemistry C*, 16688, 119, 2015.

First, E.L., and Floudas, C.A., *Microporous and Mesoporous Materials*, 32, 165, 2013.

First, E.L., Gounaris, C.E., Wei, J., and Floudas, C.A., *Physical Chemistry Chemical Physics*, 17339, 13, 2011.

Flor, R.S., and Alan, L.M., Mixed-Gas Adsorption, *AIChE Journal*, 1141, 47, 2001.

Fowkes, F.M., *J. Phys. Chem.*, 67, 2538, 1963.

Fowkes, F.M., *Ind. Eng. Chem.*, 12, 40, 1964.

Fowkes, F.M., *J. Phys. Chem.*, 84, 510, 1980.

Fowkes, F.M., *J. Adhesion Sci. Technol.*, 1, 7, 1987.

Fragneto, G., Su, T.J., Lu, J.R., Thomas, R.K., and Rennie, A.R., *Physical Chemistry Chemical Physics*, 5214, 2, 2000.

Franks, F., and Ives, D.J.C., *Q. Env. Chem., Soc.*, 20, 1, 1966.

Freundlich, H., *Colloid and Capillary Chemistry*, Methuen, London, 1926.

Fu, D., Lu, J-F., Bao, T-Z., and Li, Y-G., *Industrial & Engineering Chemistry Research*, 320, 39, 2000.

Gaines, G.L., Jr., *Insoluble Monolayers at Liquid-Gas Interfaces*, Wiley-Interscience, New York, 1966.

Gamba, O., Hulva, J., Pavelec, J., Bliem, R., Schmid, M., Diebold, U., and Parkinson, G.S., *Topics in Catalysis*, 11, 1, 2016.

Gamba, O., Novi, H., Pavelec, J., Bliem, R., Schmid, M., Diebold, U., Stierle, A., and Parkinson, G.S., *The Journal of Physical Chemistry C*, 20459, 119, 2015.

Gannon, M.G.J., *Faber Philos. Mag.*, A37, 117, 1978.

Gao, C., Lee, J.W., Spivey, J.P., Paper SPE 29173, SPE Eastern Reg Conference, Charleston, WV, 1994.

Garrone, E., Ghiotti, G., Fubini, B., *J. Chem. Soc. Faraday Trans.*, 2613, 77, 1981.

GCCSI, *The Global Status of CCS 2010*, Global CCS Institute, Canberra, 2010.

GCCSI (Global CCS Institute), *Enid Fertilizer CO2-EOR Project*, 2016a, www.globalccsin stitute.com/projects/enid-fertilizer-co2-eor-project (accessed 25 August 2016).

GCCSI, *Snøhvit CO2 Storage Project*, Snøhvit-co2-storage-project, 2016b.

GCCSI, *Global Storage Portfolio: A Global Assessment of the Geological CO2 Storage Resource Potential*, Global CCS Institute, Melbourne, 2016c.

GCCSI, *Global Status of CCS: Special Report, Understanding Industrial CCS Hubs and Clusters*, Global CCS Institute, Melbourne, 2016d.

GCCSI and Société Générale, *Targeted Report: Financing Large Scale Integrated CCS Demonstration Projects*, Global CCS Institute, Melbourne, 2014.

Gill, S.J., Dee, S.F., Olofsson, G., and Wadso, I., *J. Phys. Chem.*, 89, 3758, 1985.

Good, R.J., *Wetting*, C.J. van Oss, Ed., Plenum Press, New York, 1992.

Good Plant Design and Operation for Onshore Carbon Capture Installations and Onshore Pipelines—5 Carbon Dioxide Plant Design, Energy Institute, London, 2012.

Green, H.S., *The Molecular Theory of Fluids*, Dover Publications, New York, 1970.

Gregg, S.J., *The Surface Chemistry of Solids*, Chapman & Hall, London, 1961.

Gregg, S.J., and Sing, K.S.W., *Adsorption, Surface Area and Porosity*, Academic Press, London, 1982.

Gumma, S., and Talu, O., *Adsorption*, 17, 9, 2003.

Gumma, S., and Talu, O., Net Adsorption: A Thermodynamic Framework for Supercritical Gas Adsorption and Storage in Porous Solids, *Langmuir*, 17013, 26, 2010.

Guo, B., Chang', L., and Xiel, K., *Journal of Natural Gas Chemistry*, 223, 15, 2006.

Guo, H.X., *Applied Chemical Engineering Kinetics*, Vol. 17, Chem Ind Press, Beijing, 2003.

Hall, W.T., *Analytical Chemistry*, John Wiley, New York, 1937.

Haq, S., and Hodgson, A., *The Journal of Physical Chemistry C*, 5946, 111, 2007.

Harkins, R.W., and Wamper, J., *Am. Chem. Soc.*, 53, 850, 1931.

Harkins, W.D., *The Physical Chemistry of Surface Films*, Reinhold, New York, 1952.

Harlick, P.J.E., and Tezel, F.H., *Microporous Macroporous Matter*, 71, 79, 2004.

Harte, W.E., and Anderson, E.T., *Proc. Natl. Acad. Sci.*, 8864, 87, 1990.

Haynes, W.M. (ed.), *Handbook of Chemistry and Physics*, 95th Edition, CRC Press, Boca Raton, FL, 2014.

He, Y., Luo, J., Li, Y., Jia, H., Wang, F., Zou, C., and Zheng, C., *Energy Fuels*, 11404, 31, 2017.

He, Y., Shi, S., Ahn, J., Kang, W., Lee, C-H., *International Journal of Coal Geology*, 145, 77, 2009.

Hench, L.L., and Wilson, J., *Introduction to Bioceramics*, World Scientific, Singapore, 1993.

Heinze, C., Meyer, S., Goris, N., Anderson, L., Steinfeldt, R., Chang, N., Le, C., and Bakker, D.C.E., *Earth Sym Dynamics*, 6, 327, 2015.

Henderson, J.R., *Mol. Phys.*, 39, 709, 1980.

Herman, R.B., *J. Phys. Chem.*, 76, 2754, 1972.

Hernáinz, F., and Caro, A., *Journal of Chemical & Engineering Data*, 46, 107, 2001.

Herzfeld, H., *Science*, 256, 88, 1992.

Herzog, H., *Envirn Science & Tech.*, 148, 35, 2001.

Herzog, H., *Lessons Learned from CCS Demonstration and Large Pilot Projects*, An MIT Energy Initiative Working Paper, Massachusetts Institute of Technology, Cambridge, MA, 2016.

Herzog, H., Drake, E., and Adams, E., *MIT, Energy Lab. Report, DOE*, 1997.

Hill, T.L., Theory of Physical Adsorption, *Adv. Catal.*, 211, 1952.

Hinkov, I., Lamari, F.D., Langlois, P., Dickson, M., Chile's, C., Pentchev, I., *J Chemical Technology & Metallurgy*, 600, 51, 2016.

Hoar, T.P., and Mellord, D.A., *Trans. Faraday Sci.*, 315, 53, 1957.

Hock, R., *Wien Ber.*, 108(AII), 1516, 1899.

Holmberg, K. (ed.), *Handbook of Applied Surface and Colloid Chemistry*, Vols. 1 and 2, John Wiley & Sons, 2002.

Holmberg, K., Jonsson, B., Kronberg, B., and Lindman, B., *Surfactants and Polymers in Aqueous Solution*, 2nd Edition, John Wiley & Sons, 2003.

Horch, S., Zeppenfeld, P., and Comsa, G., *Surface Science 331*, Part B, 908, 1995.

Idem, R. et al., *International Journal of Greenhouse Gas Control*, Elsevier, Amsterdam, 6, 40, 2015.

IEA, *World Energy Outlook 2011*, OECD/IEA, Paris, 2011.

IEA (a), *Energy Technology Perspectives 2012*, OECD/IEA, Paris, 2012a.

IEA, *Technology Roadmap: Bioenergy for Heat and Power*, OECD/IEA, Paris, 2012b.

IEA, *Technology Roadmap: Carbon Capture and Storage*, OECD/IEA, Paris, 2013.

IEA, *CCS 2014—What Lies in Store for CCS?* OECD/IEA, Paris, 2014a.

IEA, *Energy, Climate Change and Environment, 2014 Insights*, OECD/IEA, Paris, 2014b.

IEA, *World Energy Investment Outlook Special Report*, OECD/IEA, Paris, 2014c.

IEA, *Energy Technology Perspectives 2015*, OECD/IEA, Paris, 2015a.

IEA, *World Energy Outlook 2015*, OECD/IEA, Paris, 2015b.

IEA, *World Energy Outlook Special Report: Energy and Climate Change*, OECD/IEA, Paris, 2015c.

IEA, *Storing CO_2 Through Enhanced Oil Recovery*, OECD/IEA, Paris, 2015d.

IEA, *Energy Technology Perspectives*, OECD/IEA, Paris, 2016a.

IEA, *Tracking Clean Energy Progress*, Energy Technology Perspectives, 2016b.

IEA, *Energy, Climate Change and Environment, Insights*, OECD/IEA, Paris, 2016c.

IEA, *Energy, Climate Change and Environment, 2016 Insights*, OECD/IEA, Paris, 2016d.

IEA/Carbon Sequestration Leadership Forum (CSLF), *IEA/CSLF Report to the Muskoka 2010 G8 Summit, Progress and Next Steps*, OECD/IEA, Paris, 2010.

IEAGHG, *CO2 Capture in the Cement Industry*, 2008/3, IEAGHG, Cheltenham, 2008.

IEAGHG, *Global Storage Resources Gap Analysis for Policy Makers*, 2011/10, IEAGHG, Cheltenham, 2011a.

IEAGHG, *Potential for Biomass and Carbon Dioxide Capture and Storage*, 2011/06, IEAGHG, Cheltenham, 2011b.

IEAGHG, *CO2 Pipeline Infrastructure*, 2013/18, IEAGHG, Cheltenham, 2013a.

IEAGHG, *Deployment of CCS in the Cement Industry*, 2013/19, IEAGHG, Cheltenham, 2013b.

IEAGHG, *Deployment of CCS in the Cement Industry*, 2013/19, IEAGHG, Cheltenham, 2013c.

IEAGHG, 20 Years of Carbon Capture and Storage OECD/IEA 2016 Accelerating Future Deployment IEAGHG, *The Process of Developing a CO2 Test Injection: Experience to Date and Best Practice*, 2013/13, IEAGHG, Cheltenham, 2013.

IEAGHG (IEA Greenhouse Gas R&D Programme), *Integrated Carbon Capture and Storage Project at SaskPower's Boundary Dam Power Station*, 2015/06, IEAGHG, Cheltenham, 2015.

IEA Publications, 9, rue de la Fédération, 75739 Paris Cedex 15 Printed in France by IEA, November 2016.

Integrated Gasification Combined Cycle for Carbon Capture Storage Claverton Energy Group Conference 24th October Bath.

Introduction to Carbon Capture and Storage—Carbon Storage and Ocean Acidification Activity, *Commonwealth Scientific and Industrial Research Organisation (CSIRO) and the Global CCS Institute* (accessed 07 March 2013).

Ionel, I., Cebrucean, D., Popescu, F., Viorica Cebrucean, H.A.R.E.A., Dungan, L.I., and Padure, G., *Thermotechnica*, 43, 2, 2008.

Ionel, I., Oprisa Stanescu, P.D., Gruescu, C.L., Savu, A., and Ungureanu, C., *Revista de Chimie*, 57, 1306, 2006.

Ionel, I., Ungereanu, C., OprisaStanesan, P.D., and Tenchea, A., *15-th European Biomass Conference & Exhibition*, CD Proceedings, V24I4, 2007.

IPCC, *IPCC Second Assessment, Climate Change 1995*, IPCC, Cambridge University Press, Cambridge, UK and New York, 1995.

IPCC, *IPCC Special Report on Carbon Dioxide Capture and Storage*. Prepared by Working Group III of the Intergovernmental Panel on Climate Change. Metz, B., Davidson, O., de Coninck, H.C., Loos, M., and Meyer, L.A. (eds.). Cambridge University Press, Cambridge, UK and New York, 2005a.

IPCC, *Special Report on Carbon Dioxide Capture and Storage*, IPCC, Cambridge University Press, New York, 2005b.

IPCC (Intergovernmental Panel on Climate Control), Fourth Assessment Report, February 2007.

IPCC (Intergovernmental Panel on Climate Change), Carbon Storage, Report for Policymakers, Cambridge Univ. Press, New York, 2011.

IPCC (Intergovernmental Panel on Climate Change), *Climate Change 2014: Mitigation of Climate Change Summary for Policymakers: Contribution of Working Group III to the Fifth Assessment Report of the Intergovernmental Panel on Climate Change*, IPCC, Cambridge University Press, Cambridge, UK and New York, 2014a.

IPCC, *Assessing Transformation Pathways, Chapter 6, in Climate Change 2014: Mitigation of Climate Change: Contribution of Working Group III to the Fifth Assessment Report of the Intergovernmental Panel on Climate Change*, IPCC, Cambridge University Press, Cambridge, UK and New York, 2014b.

IPCC, Special Report "Land is a Critical Resource", Geneva, Switzerland, 2019.

Ishikawa, S., Hada, S., and Funasaki, N., *J. Phys. Chem.*, 11508, 99, 1995.

Jansen, H.R.S., and Sogor, L., *J. Colloid Interface Sci.*, 424, 40, 1972.

Japan CCS Co., *Progress of the Tamakomai CCS Demonstration Project*, Presentation to the Combined Meeting of the IEAGHG Modelling and Monitoring Networks, 7 July 2016.

Jimenez, E., *Journal of Chemical & Engineering Data*, 1435, 44, 1999.

Jiménez, E., Casas, H., Segade, L., and Franjo, C., *Journal of Chemical & Engineering Data*, 45, 862, 2000.

Jones, D.G. et al., *International Journal of Greenhouse Gas Control*, Elsevier, Amsterdam, 350, 40, 2015.

Joosten, G.E., and Danckwerts, P.V., *Journal of Chemical and Engineering Data*, 452, 17, 1972.

Jorgensen, W.L., *J. Phys. Chem.*, 5813, 92, 1988.

Jorgensen, W.L., Blake, J.F., Madura, J.D., and Wierschke, S.D., *Am. Chem. Soc. Symp. Ser.*, 353, 200, 1987.

Jose, P.D., *Astron, J.*, 193, 70, 1965.

Jude, A.D., Rao, M., Sircar, S., Gorte, R.J., and Myers, A.L., *Langmuir*, 4333, 13, 1997.

Junker, H. & Folmer, F., 10th Eurpean Conference Biomass for Energy & Industry, 1482, 1998.

Kamusewitz, H., Possart, W., and Paul, D., *Colloids and Surfaces A: Physicochemical and Engineering Aspects*, 156 (1), 271, 1999.

Kauzmann, W., *Chem. Rev.*, 43, 219, 1948.

Keller, J.U., Staudt, R., and Tomalla, M., *Berichte der Bunsengesellschaft fur physikalische Chemie*, 28, 96, 1992.

Kemper, J., *International Journal of Greenhouse Gas Control*, 401, 40, 2015.

Kimmel, A., Matthiesen, J., Baer, M., Mundy, C.J., Petrik, N.G., Smith, R.S., Dohnálek, Z., and Kay, B.D., *Journal of the American Chemical Society*, 131 (35), 12838–12844, 2009.

Kimura, H., and Nakano, H.J., *J. Phys. Soc. Jpn.*, 54, 1730, 1985.

Knofel, C., Martin, C., Hornebecq, V., and Llewelyn, P.L., *J. Phys. Chem. C*, 21726, 113, 2009.

Knotts, T.A., Wilding, W.V., Oscarson, J.L., and Rowley, R.L., *Journal of Chemical & Engineering Data*, 158, 46, 2001.

Kohl, A.L., and Nielson, R.B., *Gas Purification*, Gulf Publishing Company, Houston, TX, 1997.

Korotcenkov, G., *Handbook of Gas Sensor*, Springer, New York, 2013.

Kramer, E.J., *Macromolecules*, 30, 1906, 1997.

Krishna, R.M. & van Baten, J., *Sep. Purif. Technol.*, 120, 87, 2012.

Kubek, D.J., Sharp, C.R., Kuper, D.E., Clark, M.E., Dio, M.D., Whyall, M., Gasification Technology Conference, San Francisco, USA, 2002.

Kuila, U., (*Ph.D. thesis*), *Measurement and interpretation of Porous and pore size distribution in Mudrocks; Mosher, K., The impact of pore size on methane and CO_2 adsorption on carbon*, 2011. Dept. of Petr. Eng., Univ. of Colorado, Golden, 2013.

Kulkarni, A., Sholl, R., and David, S., *Industrial & Engineering Chemistry Research*, 8631, 51, 2012.

Kumar, A., *Industrial & Engineering Chemistry Research*, 38, 4135, 1999.

Kumar, A., *The Journal of Physical Chemistry B*, 104, 9505, 2000.

Kwok, D.Y., Li, D., and Neumann, A.W., *Langmuir*, 10, 1323, 1994.

Lackner, K.S., Ziock, H., and Grimes, P., *Carbon Dioxide Extraction from Air: Is It an Option?* Proceedings of the 24th Annual Technical Conference on Coal Utilization & Fuel Systems, 885, 1999.

Lang, K., *Sun, Earth and Sky*, Springer, New York, 2006.

Lange, N.A., and Forker, G.M., *Handbook of Chemistry*, 10th Edition, McGraw-Hill, New York, 1967.

Langevin, D., *Langmuir*, 3206, 16, 2000.

Langmuir, I., *Proc. Nat. Acad. Sci.*, 141, 3, 1917.

Langmuir, I., *J. Am. Chem. Soc.*, 1361, 40, 1918.

Le, T.D., and Weers, J.G., *J. Phys. Chem.*, 99, 6739, 1995.

Leckel, D., *J Air Waste Manage Assoc.*, 645, 53, 2009.

Lee, J-W., Park, S-B., and Lee, H.D., *Journal of Chemical & Engineering Data*, 45, 166, 2000.

Lehner, B., Hohage, M., and Zeppcnfeld, P., *Surface Science*, 454, 251, 2000.

Leung, D.Y.C., Caramanna, G., and Marato-Valor, M.M., *Renew. Sust. Energ. Rev.*, 426, 39, 2014.

Li, J.R., Ma, Y., McCarthy, M.C., Sculley, J., Yu, J., Hae-Jeong, K., Balbuena, P.B., and Zhou, H-C., *Coordination Chemistry Reviews*, 1791, 255, 2011.

Lightfoot, E.N., and Cockrem, M.C.M., *Separation Science Technology*, 165, 22, 1987.

Linek, V., and Benes, P., A Study of the Mechanism of Gas Absorption into oil}Water Emulsions, *Chemical Engineering Science*, 1037, 31, 19176.

Liu, S.R., Dohnfilek, Z., Smith, R.S., and Kay, B.D., *The Journal of Physical Chemistry B*, 108 (11), 3644, 2004.

Lowell, S., Shields, J.E., Thomas, M.A., and Thommes, M., *Characterization of Porous Solids and Powders: Surface Area, Pore Size, and Density*, Springer, Dordrecht, 2006.

Lyman, W.J., Reehl, W.J., and Rosenblatt, D.H. (eds.), *Handbook of Chemical Property Estimation Methods*, American Chemical Society, Washington, DC, 1990.

Maier, S., and Salmeron, M., *Accounts of Chemical Research*, 2783, 48, 2015.

Manuel, O.K., *Energy Environ.*, 131, 20, 2009.

Mareschat, M., Meyer, M., and Turq, P., *J. Phys. Chem.*, 95, 10723, 1991.

Masel, R.L., *Principles of Adsorption & Reactions on Solid Surfaces*, Wiley, New York, 1996.

Mathews, G., *Journal of Earth System Science*, 391, 118, 2009.

Matijevic, E. (ed.), *Surface and Colloid Science*, Vol. 1(9, Wiley-Interscience, New York, 1976.

Matthews, H.D., and Caldeira, K., *Geophys Res Lett*, 35, 4, 2008.

McAuliffe, C., *J. Phys. Chem.*, 70, 1267, 1966.

McCormmach, R., *ISIS*, 459, 61 (4), 1970.

McDonald, A.B., Thomas, M., Mason, J.A., Kong, X. et al., Cooperative Insertion of CO_2 in Diamine-Appended Metal-Organic Frameworks, *Nature*, 519, 303, 2015.

McHale, G., Kab, N.A., Newton, M.I., Rowan, S.M., *Journal of Colloid and Interface Science*, 186, 453, 1997.

McLachlan, D.S., Blaskzkiewicz, M., and Newnham, R.E., *J. Ceramic Soc.*, 930, 73, 1990.

McLure, I.A., Sipowska, J.T., and Pegg, I.L., *J. Chem. Therm.*, 14, 733, 1982.

McLure, I.A., Soares, V.A.M., and Williamson, A.M., *Langmuir*, 9, 2190, 1993.

Mehta, V.D., and Sharma, M.M., *Chemical Engineering Science*, 461, 26, 1971.

Menke, T.J., Funke, Z., Maier, R-D., and Kressler, J., *Macromolecules*, 33, 6120, 2000.

Metz, 12B., Davidson, O., Coninck, K., Loos, M., Meyer, L., (eds.), *IPCC: Special Report on Carbon Dioxide Capture and Storage*, Cambridge University Press, Cambridge, UK, 2005.

MIT (Massachusetts Institute of Technology), *Illinois Industrial Carbon Capture and Storage Fact Sheet: Carbon Dioxide Capture and Storage Project*, MIT, Cambridge, MA, 2016.

Moniz, E.J., and Tinker, S.W., University of Texas Austin Symposium, July 23, 2010.

Monson, P.A., *Microporous and Mesoporous Materials*, 47, 160, 2012.

Mosher, K., (*M.Sc. Thesis*), *Dept. of Energy Resources Engnn.*, Stanford University Press, Stanford, CA, 2011.

Motomura, K., *Capillarity Today, Lecture Notes in Physics*, Vol. 386, K. Christmann, Ed., Springer-Verlag, Berlin, 1990.

Mumford, N., and Philips, L., *J. Am. Chem. Soc.*, 2, 5295, 1930.

Murad, S., *Chem. Eng. Commun.*, 24, 353, 1983.

Myers, A. & Monson, P.A., *Adsorption*, 591, 20, 2014.

Myers, A.L., *AIChE Journal*, 29, 691, 1983.

Myers, A.L., *Pure & Appl. Chem.*, 61, 1949, 61, 1989.

Myers, A.L., *Adsorption*, 37, 11, 2005.

Myers, A.L., and Monson, P., Adsorption in Porous Materials at High Pressure: Theory and Experiment, *Langmuir*, 10261, 18, 2002.

Myers, A.L., and Prausnitz, J.M., Thermodynamics of Mixed-Gas Adsorption, *AIChE Journal*, 121, 11, 1965.

Myers, A.L., and Sircar, S., *J. Phys. Chem.*, 76, 3412, 1972.

Myers, R.S., and Clever, H.L., *J. Chem. Therm.*, 6, 949, 1974.

Nagai, K., and Hirashima, A., *Surface Science*, 616, 187, 1987.

National Audit Office (United Kingdom), *Briefing for the House of Commons Environmental Audit Committee*, Sustainability in the Spending Review, London, July 2016.

Navaza, J.M., *Journal of Chemical & Engineering Data*, 41, 806, 1996.

Navaza, J.M., *Journal of Chemical & Engineering Data*, 43, 128, 1998a.

Navaza, J.M., *Journal of Chemical & Engineering Data*, 43, 158, 1998b.

NETL (National Energy Technology Laboratory), *Cost and Performance Baseline for Fossil Energy Plants Supplement: Sensitivity to CO2 Capture Rate in Coal-Fired Power Plants*, DOE/NETL-2015/1720, NNETL, Pittsburgh, 2015.

NETL 2007 Carbon Sequestration Atlas, National Energy Tech. Lab., Pittsburgh, 2007.

Newman, F.H., and Searle, V.H., *The General Properties of Matter*, Butterworths Scientific, London, 1957.

Nienstedt, W., Hänninen, O., Arstila, A., and Björkqvist, S-E., *Ihmisen Fysiologia a Anatomia*, 8th Edition, WSOY, 1992.

Nordholm, S., *Langmuir*, 14, 396, 1998.

Novotny, Z., Argentero, G., Wang, Z., Schmid, M., Diebold, U., and Parkinson, G.S., *Phys Rev Lett*, 21, 108, 2012.

Novotny, Z., Mulakaluri, N., Edes, Z., Schmid, M., Pentcheva, R., Diebold, U., and Parkinson, G.S., *Physical Review B*, 19, 87, 2013.

OECD/IEA 20 Years of Carbon Capture and Storage Accelerating Future Deployment, OECD/IEA, Paris, 2016.

Oh, T., Renewable Sustainable Energy, *Rev.*, 2697, 14, 2010.

Orban, J.M., *Macromolecules*, 29, 7553, 1996.

Otero-Arean, C., Manoilova, O.V., Palomino, G.T., Delgado, M.R., Tsyganenko, A.A., Bonelli, B., Garrone, E., *Phys. Chem.* 5713, 4, 2002.

Paoletti, E., Nali, C., and Lorenzini, G., *Phyton (Austria)*, Special Issue, Global Change, 149, 42, 2002.

Park, J.Y., Yoon, S.J., and Lee, H., *Environ. Sci. Technol.*, 1670, 37, 2003.

Parkinson, G.S., Dolmfilek, Z., Smith, L.S., and Kay, B.D., *The Journal of Physical Chemistry C*, 10233, 113, 2009.

Parkinson, G.S., Manz, T.A., Novotny, Z., Sprunger, P.T., Kurtz, R.L., Schmid, M., Shell, D.S., and Diebold, U., *Physical Review B*, 85 (19), 2012.

Parkinson, G.S., Novotný, Z., Jacobson, P., Schmid, M., and Diebold, U., *Journal of the American Chemical Society*, 133 (32), 12650, 2011.

Parkinson, G.S., Mulakaluri, N., Losovyj, Y., Jacobson, P., Pentcheva, R., and Diebold, U., *Physical Review B*, 82 (12), 125413, 2010.

Parratt, I.G., *Phys. Rev.*, 95, 359, 1954.

Parsegian, V.A., *Van der Waals Forces*, Cambridge University Press, New York, 2006.

Partington, J.R., *An Advanced Treatise of Physical Chemistry*, Vol. II, Longmans, Green, New York, 1951.

Pearson, D., and Robinson, E., *J. Chem. Soc., Faraday Soc.*, 736, 1934.

Penas, A., Calvo, E., Pintos, M., Amigo, A., and Bravo, R., *Journal of Chemical & Engineering Data*, 45, 682, 2000.

Penfold, J., and Thomas, R.K., *J. Phys. Cond. Matter*, 2, 1369, 1990.

Petrova, D., Kostadinova, P., Sokolovski, E., and Dombalov, I., *Greenhouse Gases & Environmental Protection & Ecology*, 679, 3, 2006.

Petty, M.C., *Langmuir-Blodgett Films*, Cambridge University Press, Cambridge, 2004.

Phelps, J., Blackford, J., Holt, J., Polton, J., *International Journal of Greenhouse Gas Control*, 9, 2015.

Picknett, R.G., and Bexon, R., *J. Colloid Interface Sci.*, 336, 61, 1977.

Puxty, G., Rowland, R., Allport, A., Yang, Q., Bown, M., Burns, R., Maeder, M., and Attalla, M., *Environ. Sci. Technol.*, 6427, 43, 2009.

Pohorecki, R., and Moniuk, W., *Chemical Engineering Science*, 43, 1677, 43, 1988.

Preuss, M., and Butt, H-J., *J. Colloid Interface Sci.*, 208, 468, 1998.

Rabinowitch, E.G., *Photosynthesis*, Wiley, New York, 1969.

Rabo, J.A., *Zeolites Chemistry & Catalysis*, Vol. 171, ACS Monograph, Amer. Chem. Soc., Washington, DC., 1976.

Rabo, J.A., Elek, L.F., and Francis, J.N., *Studies in Surface Sci. & Catalysis*, 490, 7, 1981.

Rackley, S.A., *Carbon Capture & Storage*, Butterwort Heinemann, Burlington, 2010.

Radosz, M., Hu, X., Krutkramelis, K., and Shen, Y., *Ind. Eng. Chem. Res.*, 3783, 47, 2008.

Raj, K., Shashi, V., and Kumar, S., *Indian J. of Chemical Tech.*, 704, 11, 2004.

Ramachadran, G.N., and Sasisekharan, V., *Adv. Protein Chem.*, 23, 283, 1968.

Raschi, A., Tognetti, R., Minnocci, A, Penuelas, J., and Jones, M.B., *J. Exptal. Iology*, 1135, 347, 2000.

Reid, R.C., and Sherwood, T.K., *The Properties of Gases and Liquids: Their Estimation and Correlation*, McGraw-Hill, New York, 1996.

Reid, R.C., Sherwood, T.K., and The Properties of Gases and Liquids (Their Estimation an Ringrose, P.S., *The In Salah CO2 Storage Project: Lessons Learned and Knowledge Transfer*, Proceedings of the 11th International Conference on Greenhouse Gas Control Technologies, Energy Procedia, 6226, 37, 2013.

Rey, A.R., *Langmuir*, 16, 845, 2000.

Rigby, S.P., Hasan, M., Stevens, L., Williams, H.E.L. & Fletcher, R.S., *Industrial Engineering Chemistry Research*, 14822, 56, 2017.

Rinker, E.B. Ashour, S., and Sandall, S., *Chem Engn Sci.*, 50, 755, 1995.

Rinker, E.B., Ashour, S.S., Sandall, O.C., *Ind. Eng. Chem. Res.*, 1107, 35, 1996.

Robinson, A.B., Robinson, N.F., and Noon, W., *J. Am. Phys. Surgeons*, 12, 82, 2007.

Rochelle, G.T., *Science*, 1652, 325, 2009.

Rols, J.L., Condoret, J.S., Fonade, C., and Goma, G., *Biotechnology and Bioengineering*, 427, 35, 1990.

Romero, E., *Journal of Chemical & Engineering Data*, 42, 57, 1997.

Roser, S.J, Felici, R., and Eaglesham, A., *Langmuir*, 3853, 10, 1994.

Ross, H.E., Hagin, P., and Zoback, M.D., *Int. J. Greenhouse Con.*, 773, 3, 2009.

Ross, S., (ed.), *Chemistry & Physics of Surfaces*, American Chemical Society Publications, Washington, DC, 1971.

Rouqerol, J., Rouqerol, F., and K.S.W. Sing, *Adsorption by Powders and Porous Solids: Principles, Methodology and Applications*, Academic Press, New York, 1998.

Rowan, S.M., Newton, M.L., and McHale, G., *J. Phys. Chem.*, 99, 13268, 1995.

Rowlinson, J.S., and Widom, B., *Molecular Theory of Capillarity*, Dover, London, 2003.

Rubin, E.S., Davison, J.E., and Howard, J., and Herzog, H.J., *International Journal of Greenhouse Gas Control*, Vol. 40, Elsevier, Amsterdam, 378, 40, 2015.

Saha, K., *The Earth's Atmosphere*, Springer, New York, 2008.

Salajan, M., Glinski, J., Chavepeyer, G., and Platten, J.K., *J. Colloid Interface Sci.*, 164, 387, 1994.

Salem, M.M.K., Braeuer, P., Szombathely, M.V., Heuchel, M., Harting, P., Quitzsch, K., and Jaroniec, M., *Langmuir*, 14, 3376, 1998.

Sanz-Pérez, E.S., Murdock, C.R., Didas, S.A., and Jones, C.W., *Chem. Rev.*, 11840, 116, 2016.

Sarkisov, L., and Monson, P.A., *Langmuir*, 9857, 16, 2000.

Savitz, A., Myers, A.L. & Gorte, R.J., *J. Phys. Chem., B*, 3687,103, 1999.

SBC Energy Institute, *Low Carbon Energy Technologies Fact Book Update: Carbon Capture and Storage at a Crossroads*, Energy Institute, The Hague, 2016.

Scheer, H., *Energy Autonomy: The Economic, Social and Technological Case for Renewable Energy*, Earthscan Printing house, London, 2007.

Scherrer, P.H., Schou, J., Busch, R.I., Kosovichev, A.G., Bogart, R.S., Hoeksema, J.T., Liu, Y., Duvall, T.L ., Zhao, J., Title, A.M., Schrijver, C.J., Tarbell, T.D., Tomczyk, S., *The Solar Dynamics Observatory*, 207, 2013.

Schulz, K.G., Riebsell, U., Rost, B., Thoms, S., and Zeebe, R.E., *Marine Chemistry*, 53, 10, 2006.

Schwartz, L.W., *Langmuir*, 1859, 15, 1999.

Schwarz, S.A., *Macromolecules*, 29, 899, 1996.

Shafeeyan, M.S., Ashri, W.M., Amirhossein, W.D., and Shamiri, H.A, *Chem Eng Resr & Design*, 961, 92, 2014.

Sharma, M.M., and Danckwerts, P.V., *Transactions of the Faraday Society*, 59, 386, 1963.

Sheehan, W.F., *Physical Chemistry*, 2nd Edition, Allyn & Bacon, Boston, 1970.

Sheludko, A., *Colloid Chemistry*, Elsevier Pub. Co., New York, 1966.

Shi, X.D., Brenner, M.P., and Nagel, S.R., *Science*, 265, 219, 1994.

Silin, D., and Kneafsey, T. *Journal of Canadian Petroleum Technology*, 464, 51, 2012.

Sing, K.S.W., Everett, D.H., Haul, R.A.W., Moscou, L., Pierrotti, R.A., Rouquerol, J., and Siemienewska, T., *Pure Appl. Chem.*, 57, 603, 1985.

Sircar, S., *AIChE Journal*, 1169, 47, 2001.

Smith, J.M., Van Ness, H.C., and Abbott, M.M., *Introduction to Chemical Engineering Thermodynamics*, 5th Edition, McGraw-Hill, New York, 1996.

Smith, L.S., Matthiesen, J., and Kay, B.D., *The Journal of Physical Chemistry A*, 8242, 118, 2014.

Smith, R.S., Li, Z., Chen, L., Dohnfilek, Z., and Kay, B.D., *The Journal of Physical Chemistry B*, 8054, 118, 2014.

Smith, R.S., May, R.A., and Kay, B.D., *The Journal of Physical Chemistry B*, 1979, 120, 2016.

Somasundaran, P. (ed.), *Encyclopedia of Surface and Colloid Science*, 2nd Edition, CRC Press, Boca Raton, 2006.

Somyajulu, G.R., *Int. J. Thermophys.*, 9, 559, 1988.

Song, A., Skibimki, E.S., DeBenedetti, W.J.I., Ortoll-Bloch, A.G., and Hines, M.A., *The Journal of Physical Chemistry C*, 9326, 120, 2016.

Song, B., Bismarck, A., Tahhan, R., and Springer, J., *Journal of Colloid and Interface Science*, 197, 68, 1998.

Sozzani, P., Bracco, S., Comotti, A., Ferretti, L., and Simonutti, R., *Angewandte. Chemie*, 117, 324, 2005.

Specovius, J., and Findenegg, G.H., *Berichte der Bunsengesellschaft für Physikalische Chemie*, 82, 174, 1978.

Stanger, R., *International Journal of Greenhouse Gas Control*, Elsevier, Amsterdam, 55, 40, 2015.

Steele, W.A., *The interactions of gases with Solid Surfaces*, Pergamon Press, London, 1974.

Stolaroff, J.K., *Capturing CO_2 from Ambient Air: A Feasibility Assessment (PhD thesis)*, 2006.

Straub, J., and Grigull, U., *Warme- und Stoffubertragung*, 13, 241, 1980.

Stubenrauch, C., Albouy, P-A., Von Klitzing, R., and Langevin, D., *Langmuir*, 3206, 16, 2000.

Sugimoto, Y., Abe, M., Jrlinek, P., Perez, R., Morita, S., Custance, O., *Nature*, 446 (1), 64, 2007.

Sumida, K., Rogow, D.L., Mason, J.A., McDonald, T.M., Bloch, E.D., Herm, Z.R., Bae, T-H., and Long, J.R., *Chemical Reviews*, 724, 112, 2012.

Suzuki, M., *Adsorption Engineering*, Elsevier Science Pub., New York, 1990.

Takahashi, R., *Jpn. J. Appl. Phys.*, 17, 2, 1983.

Tanford, C., *The Hydrophobic Effect*, 2nd Edition, John Wiley & Sons, New York, 1980.

Teifan, W., and Boily, J-F., *Surface Sci. Reports*, 595, 71, 2016.

Thambimuthu, K., Davison, J., Grupta, M., *CO2 Capture and Reuse*, Proceedings of IPCC Workshop on CCS, Regina, 16, 31, 2002.

Thomas, J.M., and Thomas, W.J., *Principles and Practice of Heterogeneous Catalysis*, VCH, Weinheim, Germany, 1997.

Tim, T., Gary, C., and Rochelle, T., *Chem. Eng. Sci.*, 3619, 59, 2004.

Timmerman, J., *Physico-Chemical Constants of Pure Organic Compounds*, Elsevier, New York, 1950. Selected Properties of Hydrocarbons and Related Compounds, Vol. 1, API Project 44, Thermodynamics Research Center, Texas, 1966.

Tovbin, Y.K., *The Molecular Theory of Adsorption in Porous Solids*, CRC Press, New York, 2017.

Tsierkezos, N.G., Kelarakis, A.E., and Molinou, I.E., *Journal of Chemical & Engineering Data*, 45, 776, 2000.

Vadgama, P. (ed.), *Surfaces and Interfaces for Biomaterials*, Woodhead, 2005.

Valenzuela, D.P., and Myers, A.L., *Adsorption Equilibrium Data Handbook*, Prentice-Hall, 261, 86, 1989.

van Gunsteren, W.F., and Berendsen, H.J.C., *Molecular Dynamics and Protein Structure*, J. Hermans, Ed., Polycrystal, Western Springs, IL, 1985.

van Oss, C.J., Good, R.J., and Chaudhury, M.K., *Langmuir*, 884, 4, 1988.

Van Santen, R.A., vanLeeuwen, P.W.N.M., Moulijn, J.A., and Averill, B.A., *Catalysis an Integrated Approach: Studies in Surface Science and Catalysis*, Vol. 123, Elsevier, Amsterdam, 1999.

Vargas, D.P., Giraldo, L., and Moreno-Pirajan, J.C., *Int. J. Mol. Sci.*, 8388, 13, 2012.

Vartuli, J.C., Roth, W.J., Leonowicz, M., Kresge, C.T., Schmitt, K.D., Chu, C. T-W., Olson, D.H., and Sheppard, E.W., *J. Am. Chem. Soc.*, 10834, 114, 1992.

Versteeg, G.F., and van Swaaij, W.P.M., *Journal of Chemical and Engineering Data*, 32, 29, 32, 1988.

Vogelsberger, W., Sonnefeld, J., and Rudakoff, G., *Z. Phys. Chemie, Leipzig*, 266, 225, 1985.

Vuong, T., and Monson, P.A., *Langmuir*, 5425, 12, 1996.

Wallqvist, A., and Covell, D.G., *J. Phys. Chem.*, 13118, 99, 1995.

Walton, K.S., Millward, A.R., Dubbeldam, D., Frost, H., John, J., Low, J.L., Yaghi, O.M., and Randall, Q.S., *J. Am. Chem. Soc.*, 406, 130, 2008.

Wang, B., Adrien, P., Côté, A.P., Furukawa, H., O'Keeffe, M., and Yaghi, O.M., *Nature*, 207, 453, 08 May 2008.

Watts, L.A., *Solvent Systems and Experimental Data on Thermal Conductivity and Surface Tension*, G.G. Aseyev, Ed., Begell House, Inc., New York, 1998.

Weisenberger, S., and Schumpe, A., *AIChE. Journal*, 298, 42, 1996.

Widdra, W., Trischberger, P., Friel, W., Menzel, D., Payne, S., and Kreuzer, H., *Physical Review B*, 4111, 57, 1998.

Wilcox, J., *Carbon Capture*, Springer, New York, 2012.

Willard, N.P., *Langmuir*, 14, 5907, 1998.

Winkler, A., Pogainer, G., and Rendulic, K.D., *Surface Science*, 886, 251, 1991.

Wooley, R.J., *Chem. Eng.*, 109, March 1986.

Wu, W., Giese, R.F., and van Oss, C.J., *Langmuir*, 11, 379, 1995.

Xiang, S., He, Y., Zhang, Z., Krishna, R., and Chen, B., *Nature Communications*, 3, 954, 2012.

Xu, X., Song, C., Bruce, G.M., Scaroni, AW., *Fuel Processing Technology*, 86, Arturo, 1457, 2005a.

Xu, X., Song, C., Miller, B.G., Scaroni, A.W., *Fuel Processing Technology*, 14, 86, 2005b.

Yalkowsky, S.H., Anik, S.T., and Valvani, S.C.J., *Phys. Chem.*, 2230, 79, 1975.

Yang, R.T., *Gas Separation by Adsorption Processes*, Butterworts Press, Boston, 1987.

Yang, R.T., *Adsorbents (Fundamentals & Applications)*, John Wiley & Sons, Hoboken, NJ, 2003.

Yang, X., Rees, R.J., Conway, W., Puxty, G., Yang, Q., and Winkler, A., *Chem. Rev.*, 9524, 117, 2017.

Yarranton, H.W., and Masliyah, J.H., *J. Phys. Chem.*, 1786, 100, 1996.

Yong, Z., and Rodrigues, A.E., *Adsorption*, 41, 7, 2001.

Young, D. M. & Crowell, A. D., *Physical Adsorption of Gases*, Butter-worth, Washington,DC, 4245, 1962.

Yu, C.H., Huang, C.H., and Tan, C.S., *Aerosols & Air Qualities*, 745, 12, 2012.

Yu, W., and Sepehrnoori, K., *Fuel*, 455, 116, 2014a.

Yu, W., and Sepehrnoori, K., *J. Can Petrol. Technol.*, 109, 53, 2014b.

Yoshida, F., Yamane, T., and Miyamoto, Y., *Industrial and Engineering Chemistry: Process Design and Development*, 570, 9, 1970.

Zecchina, A., Scarano, D., Bordiga, S., Spoto, G., and Lamberti, C., *Adv. Catal.*, 265, 46, 2001.

ZEP (European Technology Platform for Zero Emission Fossil Fuel Power Plants) and EBTP (European Biofuels Technology Platform), *Biomass with CO2 Capture and Storage: The Way Forward for Europe*, ZEP/EBTP, Brussels, 2012.

Zeppenfeld, P., *Surface and Interface Science*, 127, 2016.

Zhai, H., Ou, Y., and Rubin, E., *Environmental Science and Technology*, Vol. 49, American Chemical Society, Washington, DC, 7571, 49, 2015.

Zhang, Y., and Radovic, L.R., *Carbon*, 1867, 42, 2004.

Zou, Y., and Rodrigues, A.E., *Adsorption Science & Tech.*, 255, 19, 2001.

Index

Printed in the United States
by Baker & Taylor Publisher Services